UNCLE TUNGSTEN

UNCLE TUNGSTEN
Memories of a Chemical Boyhood

OLIVER SACKS

ALFRED A. KNOPF

NEW YORK · TORONTO

·2001

3 1460 00088 3081

Library of Congress Cataloging-in-Publication Data
Sacks, Oliver W.
Uncle Tungsten : memories of a chemical boyhood /
Oliver Sacks. — 1st ed.
p.   cm.
ISBN 0-375-40448-1
1. Sacks, Oliver W.   2. Neurologists—England—Biography.
I. Title
RC339.52.S23 A3 2001
616.8'092—dc21
[B]   2001033738

National Library of Canada Cataloguing in Publication Data
Sacks, Oliver, 1933–
Uncle Tungsten: memories of a chemical boyhood
ISBN 0-676-97261-6
1. Sacks, Oliver, 1933–   —Childhood and youth.
2. Sacks, Oliver, 1933–   —Knowledge—Chemistry.   I. Title.
QD22.S23A3 2001   540'.92   C2001-930571-0

Manufactured in the United States of America
First Edition

for Roald

CONTENTS

CONTENTS

The Periodic Table of the Elements is on pages 192–193.
A photographic insert follows page 150.

UNCLE TUNGSTEN

UNCLE TUNGSTEN

Many of my childhood memories are of metals: these seemed to exert a power on me from the start. They stood out, conspicuous against the heterogeneousness of the world, by their shining, gleaming quality, their silveriness, their smoothness and weight. They seemed cool to the touch, and they rang when they were struck.

I loved the yellowness, the heaviness, of gold. My mother would take the wedding ring from her finger and let me handle it for a while, as she told me of its inviolacy, how it never tarnished. "Feel how heavy it is," she would add. "It's even heavier than lead." I knew what lead was, for I had handled the heavy, soft piping the plumber had left one year. Gold was soft, too, my mother told me, so it was usually combined with another metal to make it harder.

It was the same with copper—people mixed it with tin to

produce bronze. Bronze!—the very word was like a trumpet to me, for battle was the brave clash of bronze upon bronze, bronze spears on bronze shields, the great shield of Achilles. Or you could alloy copper with zinc, my mother said, to produce brass. All of us—my mother, my brothers, and I—had our own brass menorahs for Hanukkah. (My father had a silver one.)

I knew copper, the shiny rose color of the great copper cauldron in our kitchen—it was taken down only once a year, when the quinces and crab apples were ripe in the garden and my mother would stew them to make jelly.

I knew zinc: the dull, slightly bluish birdbath in the garden was made of zinc; and tin, from the heavy tinfoil in which sandwiches were wrapped for a picnic. My mother showed me that when tin or zinc was bent it uttered a special "cry." "It's due to deformation of the crystal structure," she said, forgetting that I was five, and could not understand her—and yet her words fascinated me, made me want to know more.

There was an enormous cast-iron lawn roller out in the garden—it weighed five hundred pounds, my father said. We, as children, could hardly budge it, but he was immensely strong and could lift it off the ground. It was always slightly rusty, and this bothered me, for the rust flaked off, leaving little cavities and scabs, and I was afraid the whole roller might corrode and fall apart one day, reduced to a mass of red dust and flakes. I needed to think of metals as stable, like gold—able to stave off the losses and ravages of time.

I would sometimes beg my mother to take out her engagement ring and show me the diamond in it. It flashed like nothing I had ever seen, almost as if it gave out more light than it took in. She would show me how easily it scratched glass, and then tell me to put it to my lips. It was strangely, startlingly cold; metals felt cool to the touch, but the diamond was icy. That was because it conducted heat so well, she said—better than any metal—so it drew the body heat away from one's lips when they touched it. This was a feeling I was never to forget. Another time, she showed me how if one touched a diamond to a cube of ice, it would draw heat from one's hand into the ice and cut straight through it as if it were butter. My mother told me that diamond was a special form of carbon, like the coal we used in every room in winter. I was puzzled by this—how could black, flaky, opaque coal be the same as the hard, transparent gemstone in her ring?

I loved light, especially the lighting of the shabbas candles on Friday nights, when my mother would murmur a prayer as she lit them. I was not allowed to touch them once they were lit—they were sacred, I was told, their flames were holy, not to be fiddled with. I was mesmerized by the little cone of blue flame at the candle's center—why was it blue? Our house had coal fires, and I would often gaze into the heart of a fire, watching it go from a dim red glow to orange, to yellow, and then I would blow on it with the bellows until it glowed almost white-hot. If it got hot enough, I wondered, would it blaze blue, be blue-hot?

Did the sun and stars burn in the same way? Why did they never go out? What were they made of? I was reassured when I learned that the core of the earth consisted of a great ball of iron—this sounded solid, something one could depend on. And I was pleased when I was told that we ourselves were made of the very same elements as composed the sun and stars, that some of my atoms might once have been in a distant star. But it frightened me too, made me feel that my atoms were only on loan and might fly apart at any time, fly away like the fine talcum powder I saw in the bathroom.

I badgered my parents constantly with questions. Where did color come from? Why did my mother use the platinum loop that hung above the stove to cause the gas burner to catch fire? What happened to the sugar when one stirred it into the tea?

Where did it go? Why did water bubble when it boiled? (I liked to watch water set to boil on the stove, to see it quivering with heat before it burst into bubbles.)

My mother showed me other wonders. She had a necklace of polished yellow pieces of amber, and she showed me how, when she rubbed them, tiny pieces of paper would fly up and stick to them. Or she would put the electrified amber against my ear, and I would hear and feel a tiny snap, a spark.

My two older brothers Marcus and David, nine and ten years older than I, were fond of magnets and enjoyed demonstrating these to me, drawing the magnet beneath a piece of paper on which were strewn powdery iron filings. I never tired of the remarkable patterns that rayed out from the poles of the magnet. "Those are lines of force," Marcus explained to me—but I was none the wiser.

Then there was the crystal radio my brother Michael gave me, which I played with in bed, jiggling the wire on the crystal until I got a station loud and clear. And the luminous clocks—the house was full of them, because my uncle Abe had been a pioneer in the development of luminous paints. These, too, like my crystal radio, I would take under the bedclothes at night, into my private, secret vault, and they would light up my cavern of sheets with an eerie, greenish light.

All these things—the rubbed amber, the magnets, the crystal radio, the clock dials with their tireless coruscations—gave me a sense of invisible rays and forces, a sense that beneath the familiar, visible world of colors and appearances there lay a dark, hidden world of mysterious laws and phenomena.

Whenever we had "a fuse," my father would climb up to the porcelain fusebox high on the kitchen wall, identify the fused fuse, now reduced to a melted blob, and replace it with a new fuse of an odd, soft wire. It was difficult to imagine that a metal could melt—could a fuse really be made from the same material as a lawn roller or a tin can?

The fuses were made of a special alloy, my father told me, a combination of tin and lead and other metals. All of these had relatively low melting points, but the melting point of their alloy was lower still. How could this be so, I wondered? What was the secret of this new metal's strangely low melting point?

For that matter, what was electricity, and how did it flow? Was it a sort of fluid like heat, which could also be conducted? Why did it flow through the metal but not the porcelain? This, too, called for explanation.

My questions were endless, and touched on everything, though they tended to circle around, again and again, to my obsession, the metals. Why were they shiny? Why smooth? Why cool? Why hard? Why heavy? Why did they bend, not break? Why did they ring? Why could two soft metals like zinc and copper, or tin and copper, combine to produce a harder metal? What gave gold its goldness, and why did it never tarnish? My mother was patient, for the most part, and tried to explain, but eventually, when I exhausted her patience, she would say, "That's all I can tell you—you'll have to quiz Uncle Dave to learn more."

We had called him Uncle Tungsten for as long as I could remember, because he manufactured lightbulbs with filaments of fine tungsten wire. His firm was called Tungstalite, and I often visited him in the old factory in Farringdon and watched him at work, in a wing collar, with his shirtsleeves rolled up. The heavy, dark tungsten powder would be pressed, hammered, sintered at red heat, then drawn into finer and finer wire for the filaments. Uncle's hands were seamed with the black powder, beyond the power of any washing to get out (he would have to have the whole thickness of epidermis removed, and even this, one suspected, would not have been enough). After thirty years of working with tungsten, I imagined, the heavy element was in his lungs and bones, in every vessel and viscus, every tissue of his body. I thought of this as a wonder, not a curse—his body invigorated and fortified by the mighty element, given a strength and enduringness almost more than human.

Whenever I visited the factory, he would take me around the machines, or have his foreman do so. (The foreman was a short, muscular man, a Popeye with enormous forearms, a palpable testament to the benefits of working with tungsten.) I never tired of the ingenious machines, always beautifully clean and sleek and oiled, or the furnace where the black powder was compacted from a powdery incoherence into dense, hard bars with a grey sheen.

During my visits to the factory, and sometimes at home, Uncle Dave would teach me about metals with little experiments. I knew that mercury, that strange liquid metal, was incredibly heavy and dense. Even lead floated on it, as my uncle showed me by floating a lead bullet in a bowl of quicksilver. But then he pulled out a small grey bar from his pocket, and to my amazement, this sank immediately to the bottom. That, he said, was *his* metal, tungsten.

Uncle loved the density of the tungsten he made, and its refractoriness, its great chemical stability. He loved to handle it—the wire, the powder, but the massy little bars and ingots most of all. He caressed them, balanced them (tenderly, it seemed to me) in his hands. "Feel it, Oliver," he would say, thrusting a bar at me. "Nothing in the world feels like sintered tungsten." He would tap the little bars and they would emit a deep clink. "The sound of tungsten," Uncle Dave would say, "nothing like it." I did not know whether this was true, but I never questioned it.

As the youngest of almost the youngest (I was the last of four, and my mother the sixteenth of eighteen), I was born almost a hundred years after my maternal grandfather and never knew him. He was born Mordechai Fredkin, in 1837, in a small village in Russia. As a youth he managed to avoid being impressed into the Cossack army and fled Russia using the passport of a dead man named Landau; he was just sixteen. As Marcus Landau, he made his way to Paris and then Frankfurt, where he married (his wife was sixteen too). Two years later, in 1855, now with the first of their children, they moved to England.

My mother's father was, by all accounts, a man drawn equally to the spiritual and the physical. He was by profession a boot and shoe manufacturer, a *shochet* (a kosher slaughterer), and later a grocer—but he was also a Hebrew scholar, a mystic, an amateur mathematician, and an inventor. He had a wide-ranging mind: he published a newspaper, the *Jewish Standard,* in his basement, from 1888 to 1891; he was interested in the new science of aeronautics and corresponded with the Wright brothers, who paid him a visit when they came to London in the early 1900s (some of my uncles could still remember this). He had a passion, my aunts and uncles told me, for intricate arithmetical calculations, which he would do in his head while lying in the bath. But he was drawn above all to the invention of lamps—safety lamps for mines, carriage lamps, streetlamps—and he patented many of these in the 1870s.

A polymath and autodidact himself, Grandfather was passionately keen on education—and, most especially, a scientific education—for all his children, for his nine daughters no less than his nine sons. Whether it was this or the sharing of his own passionate enthusiasms, seven of his sons were eventually drawn to mathematics and the physical sciences, as he was. His daughters, by contrast, were by and large drawn to the human sciences—to biology, to medicine, to education and sociology. Two of them founded schools. Two others were teachers. My mother was at first torn between the physical and the human sciences: she was particularly attracted to chemistry as a girl (her older brother Mick had just begun a career as a chemist), but later became an anatomist and surgeon. She never lost her love of, her feeling for, the physical sciences, nor the desire to go beneath the surfaces of things, to explain. Thus the thousand and one questions I asked as a child were seldom met by impatient or peremptory answers, but careful ones which enthralled me (though they were often above my head). I was encouraged from the start to interrogate, to investigate.

Given all my aunts and uncles (and a couple more on my father's side), my cousins numbered almost a hundred; and since the family, for the most part, was centered in London (though there were far-flung American, Continental, and South African branches), we would all meet frequently, tribally, on family occasions. This sense of extended family was one I knew and enjoyed as far back as memory goes, and it went with a sense that it was our business, the family business, to ask questions, to be "scientific," just as we were Jewish or English. I was among the youngest of the cousins—I had cousins in South Africa who were forty-five years my senior—and some of these cousins were already practicing scientists or mathematicians; others, only a little older than myself, were already in love with science. One cousin was a young physics teacher; three were reading chemistry at university; and one, a precocious fifteen-year-old, was showing great mathematical promise. All of us, I could not help imagining, had a bit of the old man in us.

2

" 3 7 "

I grew up just before the Second World War in a huge, rambling Edwardian house in northwest London. Being a corner house, at the junction of Mapesbury and Exeter Roads, number 37 Mapesbury Road faced onto both, and was larger than its neighbors. The house was basically square, almost cubical, but with a front porch that jutted out, V-shaped at the top, like the entry to a church. There were bow windows that also protruded on each side, with recesses in between, and thus the roof had a most complex shape, resembling, to my eyes, nothing so much as a giant crystal. The house was built of red brick of a peculiarly soft, dusky color. I imagined this, after I learned some geology, as being old red sandstone from the Devonian age, a thought encouraged by the fact that all the roads around us—Exeter, Teignmouth, Dartmouth, Dawlish—themselves had Devonian names.

There were double front doors, with a little vestibule between them, and these led onto a hall, and thence to a passage that led back toward the kitchen; the hall and the passage had a floor of tesselated colored stones. To the right of the hall, as one entered, the staircase curved upward, its heavy bannister polished smooth by my brothers sliding down it.

Certain rooms in the house had a magical or sacred quality, perhaps my parents' surgery (both of them were physicians) above all, with its bottles of medicine, its balance for weighing out powders, the racks of test tubes and beakers, the spirit lamp, and the examining table. There were all sorts of medicines, lotions, and elixirs in a large cabinet—it looked like an old-fashioned chemist's shop in miniature—there was a microscope, and bottles of reagents for testing patients' urine, like the bright blue Fehling's solution, which turned yellow when there was sugar in the urine.

It was from this special room, where patients were admitted, but not (unless the door was left unlocked) my childish self, that I sometimes saw a glow of violet light coming out under the door and smelled a strange, seaside smell, which I later learned was ozone—this was the old ultraviolet lamp at work. I was not too sure, as a child, what doctors "did," and glimpses of catheters and bougies in their kidney dishes, retractors and speculums, rubber gloves, catgut thread and forceps—all this, I think, rather frightened me, though it fascinated me too. Once, when the door was left accidentally open, I saw a patient with her legs up in stir-rups (in what I later learned was the "lithotomy position"). My mother's obstetric bag and anesthetic bag were always ready to be grabbed in an emergency, and I knew when they would be needed, for I would hear comments like, "She's half-a-crown dilated"—comments which by their unintelligibility and mys-teriousness (were they a sort of code?) stimulated my imagination in all sorts of ways.

Another sacred room was the library, which, in the evenings at least, was especially my father's domain. One section of the library wall was covered with his Hebrew books, but there were books on every subject — my mother's books (she was fond of novels and biographies), my brothers' books, and books inherited from grandparents. One bookcase was entirely devoted to plays — my parents, who had met as fellow enthusiasts in a medical students' Ibsen society, still went to the theater every Thursday.

The library was not only for reading; on weekends, the books that were out on the reading table would be put to one side to make room for games of various sorts. While my three older brothers might be playing an intense game of cards or chess, I would play a simple game, Ludo, with Auntie Birdie, my mother's older sister, who lived with us — in my early years, she was more a play companion than my brothers were. Extreme passions developed over Monopoly, and even before I learned to play it, the prices and colors of the properties became engraved on my mind. (To this day I see the Old Kent Road and Whitechapel as cheap, mauve properties, the pale blue Angel and Euston Road next to them as scarcely any better. By contrast, the West End is clothed for me in rich, costly colors: Fleet Street scarlet, Piccadilly yellow, the green of Bond Street, and the dark, Bentley-colored blue of Park Lane and Mayfair.) Sometimes we would all join in a game of Ping-Pong, or some woodworking, using the big library table. But after a weekend of frivolities, the games would be returned to the huge drawer under one of the bookcases, and the room restored to its quiet for my father's evening reading.

There was another drawer on the other side of the bookcase, a fake drawer which, for some reason, did not open, and I frequently had a fixed dream about this drawer. Like any child, I loved coins — their glitter, their weights, their different shapes and sizes — from the bright copper farthings and halfpennies and pennies to the varied silver coins (especially the tiny silver threepenny bits — one was always concealed in the suet pudding at

Christmas) to the heavy gold sovereign my father wore on his watch chain. And I had read in my children's encyclopedia about doubloons and rubles, coins with holes in them, and "pieces of eight," which I imagined to be perfect octagons. In my dream the false drawer would open to me, displaying a glittering treasure of copper and silver and gold mixed together, coins of a hundred countries and ages, including, to my delight, octagonal pieces of eight.

I especially liked crawling into the triangular cupboard under the stairs, where the special plates and cutlery for Passover were kept. The cupboard itself was shallower than the stairs, and it seemed to me that its back wall, when tapped, sounded hollow; it must have concealed, I felt, a further space behind it, a secret passageway, perhaps. I felt snug in here, in my secret hideaway — no one besides me was small enough to fit in.

Most beautiful and mysterious in my eyes was the front door, with its stained glass panels of many shapes and colors. I would place my eye behind the crimson glass and see a whole world red-stained (but with the red roofs of the houses opposite strangely pale, and clouds startlingly distinct against a blue sky now almost black). It was a completely different experience with the green glass, and the deep violet blue. Most intriguing was the yellowish green glass, for this seemed to shimmer, sometimes yellow and sometimes green, depending on where I stood and how the sun hit it.

A forbidden area was the attic, which was gigantic, since it covered the entire area of the house, and stretched up to the peaked, crystalline eaves of the roof. I was once taken up to see the attic, and then dreamed of it repeatedly, perhaps because it was forbidden after Marcus climbed up once by himself and fell through the skylight, gashing his thigh (though once, in a story-telling mood, he told me that the scar had been inflicted by a wild boar, like the scar on Odysseus' thigh).

We had meals in the breakfast room next to the kitchen; the

dining room, with its long table, was reserved for shabbas meals, festivals, and special occasions. There was a similar distinction between the lounge and the drawing room—the lounge, with its sofa and dilapidated, comfy chairs, was for general use; the drawing room, with its elegant, uncomfortable Chinese chairs and lacquered cabinets, was for large family gatherings. Aunts, uncles, and cousins in the neighborhood would walk over on Saturday afternoons, and a special silver tea service would be pulled out and small crustless sandwiches of smoked salmon and cod's roe served in the drawing room—such dainties were not served at any other time. The chandeliers in the drawing room, originally gasoliers, had been converted to electric light sometime in the 1920s (but there were still odd gas jets and fittings all over the house so that, in a pinch, we could go back to gas lighting). The drawing room also contained a huge grand piano, covered with family photos, but I preferred the soft tones of the upright piano in the lounge.

Though the house was full of music and books, it was virtually empty of paintings, engravings, or artwork of any sort; and similarly, while my parents went to theaters and concerts frequently, they never, as far as I can remember, visited an art gallery. Our synagogue had stained glass windows depicting biblical scenes, which I often gazed at in the more excruciating parts of the service. There had been, apparently, a dispute over whether such pictures were appropriate, given the interdiction of graven images, and I wondered whether this was a reason we had no art in the house. But it was rather, I soon realized, that my parents were completely indifferent to the decor of the house or its furnishings. Indeed, I later learned that when they had bought the place, in 1930, they had given my father's older sister Lina their checkbook, carte blanche, saying, "Do what you want, get what you want."

Lina's choices—fairly conventional, except for the chinoiserie in the drawing room—were neither approved nor contested; my parents accepted them without really noticing or caring. My friend Jonathan Miller, visiting the house for the first time—this was soon after the war—said it seemed like a rented house to him, there was so little evidence of personal taste or decision. I was as indifferent as my parents to the decor of the house, though I was angered and bewildered by Jonathan's comment. For, to me, 37 was full of mysteries and wonders—the stage, the mythic background, on which my life was lived.

There were coal fires in almost every room, including a porcelain coal stove, flanked by fish tiles, in the bathroom. The fire in the lounge had large copper coal scuttles to either side, bellows, and fire irons, including a slightly bent poker of steel (my eldest brother, Marcus, who was very strong, had managed to bend it, when it was almost white-hot). If an aunt or two visited, we would all gather in the lounge, and they would hitch up their skirts and stand with their backs to the fire. All of them, like my mother, were heavy smokers, and after warming themselves by the fire, they would sit on the sofa and smoke, lobbing their wet fag ends into the fire. They were, by and large, terrible shots, and the damp butts would hit the brick wall surrounding the fireplace and adhere there, disgustingly, until they finally burned away.

I have only fragmentary, brief memories of my youngest years, the years before the war, but I remember being frightened, as a child, by observing that many of my aunts and uncles had coal black tongues—would my own, I wondered, turn black when I grew up? I was greatly relieved when Auntie Len, divining my fears, told me that her tongue was not really black, that its blackness came from chewing charcoal biscuits, and that they all ate these because they had gas.

Of my Auntie Dora (who died when I was very young), I remember nothing except for the color orange—whether this was the color of her complexion or hair, or of her clothes, or whether it was the reflected color of the firelight, I have no idea. All that remains is a warm, nostalgic feeling and a peculiar fondness for orange.

My bedroom, since I was the youngest, was a tiny room connected with my parents' bedroom, and I remember that its ceiling was festooned with strange, calcareous concretions. Michael had had this room before I was born, and had been fond of flicking gelatinous spoonsful of sago—the sliminess of which he disliked—onto the ceiling, where it would adhere with a wet smack. As the sago dried, nothing but a chalky mound would remain.

There were several rooms which belonged to nobody and had no clear function; these were used to house extras of all kinds—books, games, toys, magazines, waterproofs, sports equipment. In one small room there was nothing but a Singer sewing machine with a treadle (which my mother had bought on her marriage, in 1922) and a knitting machine of an intricate (and, to my mind, beautiful) design. My mother used it to make our socks, and I loved to watch her turning the handle, to watch the glittering steel knitting needles clacking in unison and the cylinder of wool, weighted with a lead bob, descending steadily. On one occasion I distracted her as she was making a sock, and the cylinder of wool got longer and longer, until finally it hit the floor. Not knowing what to do with this yard-long cylinder of wool, she gave it to me to keep as a muff.

These extra rooms enabled my parents to accommodate relatives like Auntie Birdie and others, sometimes for long periods. The largest of them was reserved for the formidable Auntie Annie, on her rare visits from Jerusalem (thirty years after her death, this was still referred to as "Annie's room"). When Auntie Len came to visit from Delamere, she, too, had her own room, and here she would establish herself, with her books and her tea things—there was a gas ring in the room, and she would make her own tea—and when she invited me in, I felt I was entering a different world, a world of other interests and tastes, of civility, of unconditional love.

When my uncle Joe, who had been a doctor in Malaya, became a Japanese prisoner of war, his older son and daughter stayed with us. And my parents would sometimes take in refugees from Europe during the war years. So the house, though large, was never empty; it seemed, on the contrary, to house dozens of separate lives, not just the immediate family—my parents and my three brothers and myself—but itinerant uncles and aunts, the resident staff—our nanny and nurse, the cook—and the patients themselves, who would come and go.

3

EVACUATION
OF
WOMEN AND CHILDREN
FROM LONDON, Etc.

FRIDAY, 1st SEPTEMBER.
Up and Down business trains as usual.
with few exceptions.
Main Line and Suburban services will be
curtailed while evacuation is in progress
during the day.

SATURDAY & SUNDAY.
SEPTEMBER 2nd & 3rd.
The train service will be exactly the
same as on Friday.
Remember that there will be very few
Down Mid-day business trains on Saturday.

SOUTHERN RAILWAY

EXILE

In early September 1939, war broke out. It was expected that London would be heavily bombed, and parents were put under great pressure by the government to evacuate their children to safety in the countryside. Michael, five years older than I, had been going to a day school near our house, and when it was closed at the outbreak of the war one of the assistant masters there decided to reconstitute the school in the little village of Braefield. My parents (I was to realize many years later) were greatly worried about the consequences of separating a little boy—I was just six—from his family and sending him to a makeshift boarding school in the Midlands, but they felt they had no choice, and took some comfort that at least Michael and I would be together.

This, perhaps, might have worked out well—evacuation did work out reasonably well for thousands of others. But the school, as reconstituted, was a travesty of the original. Food was rationed and scarce, and our food parcels from home were looted by the matron. Our basic diet was swedes and mangel-wurzels—giant turnips and huge, coarse beetroots grown for cattle. There was a steam pudding whose revolting, suffocating smell comes back to me (as I write almost sixty years later) and sets me retching and gagging once again. The horribleness of the school was made worse for most of us by the sense that we had been abandoned by our families, left to rot in this awful place as an inexplicable pun-ishment for something we had done.

The headmaster seemed to have become unhinged by his own power. He had been decent enough, even well liked, as a teacher in London, Michael said, but at Braefield, where he took over, he had quickly become a monster. He was vicious and sadistic, and beat many of us, with relish, almost daily. "Wilfulness" was severely punished. I sometimes wondered if I was his "darling," the one selected for a maximum of punishment, but in fact many of us were so beaten we could hardly sit down for days on end. Once, when he had broken a cane on my eight-year-old bottom, he roared, "Damn you, Sacks! Look what you have made me do!" and added the cost of the cane to my bill. Bullying and cruelty, meanwhile, were rife among the boys, and great ingenuity was exercised in finding out the weak points of the smaller children and tormenting them beyond bearable limits.

But along with the horror there were sudden delights, made sharper by their rarity and contrast with the rest of life. My first winter there—the winter of 1939–40—was an exceptionally cold one, with drifting snow, higher than my head, and long glittering icicles hanging from the eaves of the church. These snowy scenes, and sometimes fantastic snow and ice forms, conveyed me in imagination to Lapland or Fairyland. To get out of the school to the surrounding fields was always a pleasure, and the freshness and whiteness and cleanness of the snow allowed a wonderful, though brief, release from the shut-in-ness, the misery, the smell of the school. Once I somehow contrived to separate myself from the other boys and our teacher and got briefly, ecstatically, "lost" among the snowdrifts—a feeling that soon turned to terror when it became clear that I really was lost, and no longer just playing. I was very happy to be found, finally, and hugged and given a mug of hot chocolate when I got back to school.

It was during the same winter that I remember finding the windowpanes of the rectory doors covered with hoarfrost, and being fascinated by the needles and crystalline forms in this, and how I could melt some of the frost with my breath and make a little peephole. One of my teachers—her name was Barbara Lines—saw my absorption and showed me the snow crystals under a pocket lens. No two were ever quite the same, she told me, and the sense of how much variation was possible within a basic hexagonal format was a revelation to me.

There was a particular tree in a field that I loved; its silhouette against the sky affected me in a strange way. I still see it, and the winding path through the fields that led to it, when my mind drifts back. The sense that nature, at least, existed outside the dominion of school was deeply reassuring.

And the vicarage, with its spacious garden, where the school was housed, the old church next door to it, and the village itself were charming, even idyllic. The villagers were kind to these obviously uprooted and unhappy young boys from London. It was here in the village that I learned to ride horses, with a strapping young woman; she sometimes hugged me when I looked miserable. (Michael had read me parts of *Gulliver's Travels,* and I sometimes thought of her as Glumdalclitch, Gulliver's giant nurse.) There was an old lady to whom I went for piano lessons, and she would make tea for me. And there was the village shop, where I would go to buy a gob-stopper and occasionally a slice of corned beef. There were even times in school which I enjoyed: making model planes of balsa wood, and a tree house with a friend, a little red-haired boy of my own age. But, overwhelmingly, I felt trapped at Braefield, without hope, without recourse, forever—and many of us, I suspect, were severely disturbed by being there.

During the four years I was at Braefield, my parents visited us at the school, but very rarely, and I have almost no memory of these visits. When, in December 1940, after nearly a year away from home, Michael and I returned to London for the Christmas holidays, I had a complex mixture of feelings: relief, anger, pleasure, apprehension. The house felt strange and different, too: our housekeeper and cook had gone, and there were strangers there, a Flemish couple who had been among the last to make their escape from Dunkirk—my parents had offered to take them in, now that the house was nearly empty, until they found a place. Only Greta, our dachshund, seemed the same, and she greeted me with yelps of welcome, rolling on her back, wriggling with joy.

There were physical changes too: the windows were all hung with heavy blackout curtains; the inner front door, with the colored glass I had loved to look through, had been blown out by a bomb blast a couple of weeks earlier; the garden, now planted with Jerusalem artichokes for the war effort, was changed almost beyond recognition; and the old gardening shed had been replaced by an Anderson shelter, an ugly, blocky building with a thick reinforced-concrete roof.

Although the Battle of Britain was over, the Blitz was still at its height. There were air raids almost every night, and the night sky would be lit up with ack-ack fire and searchlights. I remember seeing German airplanes transfixed in the roving searchlight beams as they flew in the now-darkened skies over London. It was frightening, and also thrilling for a seven-year-old—but most of all, I think, I felt glad to be away from school and at home, protected, once again.

One night, a thousand-pound bomb fell into the garden next to ours, but fortunately it failed to explode. All of us, the entire street, it seemed, crept away that night (my family to a cousin's flat)—many of us in our pajamas—walking as softly as we could (might vibration set the thing off?). The streets were pitch dark, for the blackout was in force, and we all carried electric torches dimmed with red crêpe paper. We had no idea if our houses would still be standing in the morning.

On another occasion, an incendiary bomb, a thermite bomb, fell behind our house and burned with a terrible, white-hot heat. My father had a stirrup pump, and my brothers carried pails of water to him, but water seemed useless against this infernal fire—indeed, made it burn even more furiously. There was a vicious hissing and sputtering when the water hit the white-hot metal, and meanwhile the bomb was melting its own casing and throwing blobs and jets of molten metal in all directions. The lawn was as scarred and charred as a volcanic landscape the next morning, but littered, to my delight, with beautiful gleaming shrapnel that I could show off at school after the holidays.

A curious, and shameful, episode stays in my mind from that brief period at home during the Blitz. I was very fond of Greta, our dog (I wept bitterly when she was later killed by a speeding motorbike, in 1945), but one of my first acts, that winter, was to imprison her in the freezing coalbin in the yard outside, where her pitiful whimperings and barkings could not be heard. She was missed after a while, and I was asked, we were all asked, when we had last seen her, whether we had any idea where she was. I thought of her—hungry, cold, imprisoned, perhaps dying in the coalbin outside—but said nothing. It was only toward evening that I admitted what I had done, and Greta was fetched, almost frozen, from the bin. My father was furious and gave me "a good hiding" and stood me in a corner for the rest of the day. There was no enquiry, however, as to why I had been so uncharacteristically naughty, why I had behaved so cruelly to a dog I had loved; nor, had I been asked, could I have told them. But it was surely a message, a symbolic act of some kind, trying to draw my parents' attention to *my* coalbin, Braefield, my misery and helplessness there. Even though bombs were dropping daily in London, I dreaded returning to Braefield more than I could say, and longed to stay at home with the family, to be with them, not separated, even if we all got bombed.

I had had some religious feeling, of a childish sort, in the years before the war. When my mother lit the shabbas candles, I would feel, almost physically, the Sabbath coming in, being welcomed, descending like a soft mantle over the earth. I imagined, too, that this occurred all over the universe, the Sabbath descending on far-off star systems and galaxies, enfolding them all in the peace of God.

Prayer had been a part of life. First the Sh'mah, "Hear, O Israel . . . ," then the bedtime prayer I would say every night. My mother would wait until I had cleaned my teeth and put on my pajamas, and then she would come upstairs and sit on my bed while I recited in Hebrew, *"Baruch atoh adonai . . .* Blessed art thou, O Lord our God, King of the Universe, who makest the bands of sleep to fall upon mine eyes, and slumber upon mine eyelids. . . ." It was beautiful in English, more beautiful still in Hebrew. (Hebrew, I was told, was God's actual language, though, of course, He understood every language, and even one's feelings, when one could not put them into words.) "May it be thy will, O Lord our God and God of my fathers, to suffer me to lie down in peace, and to let me rise up again. . . ." But by this point the bands of sleep (whatever they were) would be pressing heavily upon my eyes, and I rarely got any further. My mother would bend over and kiss me, and I would instantly fall asleep.

Back at Braefield there was no kiss, and I gave up my bedtime prayer, for it was inseparably associated with my mother's kiss, and now it was an intolerable reminder of her absence. The very phrases that had so warmed and comforted me, conveying God's concern and power, were now so much verbiage, if not gross deceit.

For when I was suddenly abandoned by my parents (as I saw it), my trust in them, my love for them, was rudely shaken, and with this my belief in God, too. What evidence was there, I kept asking myself, for God's existence? At Braefield, I determined on an experiment to resolve the matter decisively: I planted two rows of radishes side by side in the vegetable garden, and asked God to bless one or curse one, whichever He wished, so that I might see a clear difference between them. The two rows of radishes came up identical, and this was proof for me that no God existed. But I longed now even more for something to believe in.

As the beatings, the starvings, the tormentings continued, those of us who remained at school were driven to more and more extreme psychological measures—dehumanizing, derealizing, our chief tormentor. Sometimes, while being beaten, I would see him reduced to a gesticulating skeleton (at home I had seen radiographs, bones in a tenuous envelope of flesh). At other times, I would see him as not a being at all, but a temporary vertical collection of atoms. I would say to myself, "He's only atoms"—and, more and more, I craved a world that was "only atoms." The violence exuded by the headmaster at times seemed to contaminate the whole of living nature, so that I saw violence as the very principle of life.

What could I do, in these circumstances, other than seek a private place, a refuge where I might be alone, absorb myself without interference from others, and find some sense of stability and warmth? My situation was perhaps similar to that which Freeman Dyson describes in his autobiographical essay "To Teach or Not to Teach."

> I belonged to a small minority of boys who were lacking in physical strength and athletic prowess . . . and squeezed between the twin oppressions of [a vicious headmaster and bullying boys]. . . . We found our refuge in a territory that was equally inaccessible to our Latin-obsessed headmaster and our football-obsessed schoolmates. We found our refuge in science. . . . We learned . . . that science is a territory of freedom and friendship in the midst of tyranny and hatred.

For me, the refuge at first was in numbers. My father was a whiz at mental arithmetic, and I, too, even at the age of six, was quick with figures—and, more, in love with them. I liked numbers because they were solid, invariant; they stood unmoved in a chaotic world. There was in numbers and their relation something absolute, certain, not to be questioned, beyond doubt. (Years later, when I read *1984,* the climactic horror for me, the ultimate sign of Winston's disintegration and surrender, was his being forced, under torture, to deny that two and two is four. Even more terrible was the fact that eventually he began to doubt this in his own mind, that finally numbers failed him, too.)

I particularly loved prime numbers, the fact that they were indivisible, could not be broken down, were inalienably themselves. (I had no such confidence in myself, for I felt I was being divided, alienated, broken down, more every week.) Primes were the building blocks of all other numbers, and there must be, I felt, some meaning to them. Why did primes come when they did? Was there any pattern, any logic to their distribution? Was there any limit to them, or did they go on forever? I spent innumerable hours factoring, searching for primes, memorizing them. They afforded me many hours of absorbed, solitary play, in which I needed no one else.

I made a grid, ten by ten, of the first hundred numbers, with the primes blacked in, but I could see no pattern, no logic to their distribution. I made larger tables, increased my grids to twenty squared, thirty squared, but still could discern no obvious pattern. And yet I was convinced that there must be one.

The only real holidays I had during the war were visits to Auntie Len in Cheshire, in the midst of Delamere Forest, where she had founded the Jewish Fresh Air School for "delicate children" (these were children from working-class families in Manchester—many had asthma, some had had rickets or tuberculosis, and one or two, I suspect, looking back, were autistic). All the children here had little gardens of their own, squares of earth a couple of yards wide, bordered by stones. I wished desperately that I could go to Delamere rather than Braefield—but this was a wish I never expressed (though I wondered if my clear-sighted and loving aunt did not divine it).

Auntie Len always delighted me by showing me all sorts of botanical and mathematical pleasures. She showed me the spiral patterns on the faces of sunflowers in the garden, and suggested I count the florets in these. As I did so, she pointed out that they were arranged according to a series—1, 1, 2, 3, 5, 8, 13, 21, etc.—each number being the sum of the two that preceded it. And if one divided each number by the number that followed it (1/2, 2/3, 3/5, 5/8, etc.), one approached the number 0.618. This series, she said, was called a Fibonacci series, after an Italian mathematician who had lived centuries before. The ratio of 0.618, she added, was known as the divine proportion or the golden section, an ideal geometrical proportion often used by architects and artists.

She would take me for long, botanizing walks in the forest, where she had me look at fallen pinecones, to see that they, too, had spirals based on the golden section. She showed me horsetails growing near a stream, had me feel their stiff, jointed stems, and suggested that I measure these and plot the lengths of the successive segments as a graph. When I did so and saw that the curve flattened out, she explained that the increments were "exponential" and that this was the way growth usually occurred. These ratios, these geometric proportions, she told me, were to be found all over nature—numbers were the way the world was put together.

The association of plants, of gardens, with numbers assumed a curiously intense, symbolic form for me. I started to think in terms of a kingdom or realm of numbers, with its own geography, languages, and laws; but, even more, of a garden of numbers, a magical, secret, wonderful garden. It was a garden hidden from, inaccessible to, the bullies and the headmaster; and a garden, too, where I somehow felt welcomed and befriended. Among my friends in this garden were not only primes and Fibonacci sunflowers, but perfect numbers (such as 6 or 28, the sum of their factors, excluding themselves); Pythagorean numbers, whose square was the sum of two other squares (such as 3, 4, 5 or 5, 12, 13); and "amicable numbers" (such as 220 and 284), pairs of numbers in which the factors of each added up to the other. And my aunt had shown me that my garden of numbers was doubly magical—not just delightful and friendly, always there, but part of the plan on which the whole universe was built. Numbers, my aunt said, are the way God thinks.

Of all the objects at home, the one I missed most was my mother's clock, a beautiful old grandfather clock with a golden face showing not only the time and date, but phases of the moon and conjunctions of the planets. When I was very young, I had thought of this clock as a sort of astronomical instrument, transmitting information straight from the cosmos. Once a week my mother would open the cabinet and wind the clock, and I would watch the heavy counterweight ascending and touch (if she let me) the long metal chimes for the hours and the quarters.

I missed its chimes painfully in my four years at Braefield and sometimes dreamed of them at night and imagined myself at home, only to wake and find myself in a narrow, lumpy bed, wet, as often as not, with my own incontinence. Many of us regressed at Braefield, and we were beaten savagely when we wet or soiled our beds.

In the spring of 1943, Braefield was closed. Almost everyone had complained to their parents about the conditions at the school, and most of them had been taken away. I never complained (nor did Michael, but he had moved to Clifton College, as a thirteen-year-old, in 1941), and finally I found myself almost the only one left. I never knew what happened exactly—the headmaster disappeared, with his odious wife and child—I was simply told, at the end of the holidays, that I would not be returning to Braefield, but going to a new school instead.

St. Lawrence College (so it seemed to me) had large and venerable grounds, ancient buildings, ancient trees—it was all very fine, doubtless, but it terrified me. Braefield, for all its horrors, was at least familiar—I knew the school, I knew the village, I had a friend or two—whereas everything at St. Lawrence was strange to me, unknown.

I have curiously little memory of the term I spent there—it seems to have been so deeply repressed or forgotten that when I mentioned it recently to someone who knew me well, and who knew much about the Braefield period, she was astonished, and said I had never spoken of St. Lawrence before. My chief memories, indeed, are of the sudden lies, or jokes, or fantasies, or delusions—I scarcely know what to call them—that I generated there.

I felt particularly alone on Sunday mornings, when all the other boys went to chapel, leaving me, a little Jewish boy, alone in the school (this had not happened at Braefield, where most of the children were Jewish). One Sunday morning there was a great storm, with violent lightning and tremendous peals of thunder—one so terrifyingly loud and close that I thought for a moment the school had been struck. When the others returned from chapel, I steadfastly insisted that I had, in fact, been struck by lightning, and that the lightning had "entered" me, and lodged in my brain.

Other fictions I maintained had relation to my childhood, or rather an alternative version or fantasy of childhood. I said that I had been born in Russia (Russia was our ally at the time, and I knew that my mother's father had come from there), and I would tell long, fanciful, richly detailed stories of jolly toboggan rides, of being wrapped in furs, and of howling wolf packs pursuing our sleigh at night. I have no memory of how these stories were received, but I stuck to them.

I maintained at other times that my parents, for some reason, had thrown me out as a child, but that I had been found by a she-wolf and brought up among wolves. I had read *The Jungle Book* and knew it almost by heart, and I was able to embroider my "recollections" richly from this, telling the amazed nine-year-olds around me about Bagheera, the black panther, and Baloo, the old bear who taught me the Law, and Kaa, my snake friend with whom I swam in the river, and Hathi, the king of the jungle, who was a thousand years old.

It seems to me as I look back on this time that I was filled with daydreams and myths, and that I was uncertain, at times, about the boundaries between fantasy and reality. It seems to me I was trying to invent an identity of an absurd yet glamorous kind. I think my sense of isolation, of being uncared for and unknown, may have been even greater at St. Lawrence than it was at Braefield, where even the sadistic attentions of the headmaster could be seen as a sort of concern, even love. I think I was, perhaps, enraged with my parents, who remained blind and deaf, or inattentive, to my distress, and so was tempted to replace them with kindly, parental Russians or wolves.

When my parents visited me at midterm in 1943 (and perhaps heard of my curious fantastications and lies), they finally realized that I was close to the edge, and that they had better bring me back to London before worse befell.

4

"AN IDEAL METAL"

I returned to London in the summer of 1943, after four years of exile, a ten-year-old boy, withdrawn and disturbed in some ways, but with a passion for metals, for plants, and for numbers. Life was beginning to resume some degree of normality, despite the bomb damage everywhere, despite the rationing, the blackout, and the thin, poor paper on which books were printed. The Germans had been turned back at Stalingrad, the Allies had landed in Sicily; it might take years, but victory was now certain.

One sign of this, for me, was the fact that my father was given, through a series of intermediaries, an unheard-of thing, a banana from North Africa. None of us had seen a banana since the start of the war, and so my father divided it, sacramentally, into seven equal segments: one each for my mother and himself, one for Auntie Birdie, and one apiece for my brothers and myself. The tiny segment was placed, like a Host, on the tongue, then savored slowly as it was swallowed. Its taste was voluptuous, almost ecstatic, at once a reminder and symbol of times past and an anticipation of times to come, an earnest, a token, perhaps, that I had come home to stay.

And yet much had changed, and home itself was disconcertingly different, utterly changed in many ways from the settled, stable household there had been before the war. We were, I suppose, an average middle-class household, but such households, then, had a whole staff of helpers and servants, many of whom were central in our lives, growing up as we did with very busy and to some extent "absentee" parents. There was the senior nanny, Yay, who had been with us since Marcus's birth in 1923 (I was never certain how her name was spelled, but imagined, after I learned to read, that it was spelled "Yea"—I had read some of the Bible, and been fascinated by words like *lo* and *hark* and *yea*). Then there was Marion Jackson, my own nanny, to whom I was passionately attached—my first intelligible words (I am told) were the words of her name, each syllable pronounced with babyish slowness and care. Yay wore a nurse's headdress and uniform, which looked to me somewhat severe and forbidding, but Marion Jackson wore soft white clothes, soft as a bird's feathers, and I would nestle against them and feel utterly secure.

There was Marie, the cook-housekeeper, with her starched apron and reddened hands, and a "daily," whose name I forget, who came in to help her. Besides these four women, there was Don, the chauffeur, and the gardener, Swain, who between them handled the heavy work of the house.

Very little of this survived the war. Yay and Marion Jackson disappeared—we were all "grown up" now. The gardener and the chauffeur had gone, and my mother (now fifty) decided to drive her own car. Marie was due to come back, but never did; and in her stead Auntie Birdie did the shopping and cooking.[1]

---

[1] Only one person stayed: Miss Levy, my father's secretary. She had been with him since 1930, and though somewhat reserved and formal (it would

Physically, too, the house had changed. Coal had become scarce, like everything else in the war, and the huge boiler had been shut down. There was a small oil burner, of very limited capacity, in its stead, and many of the extra rooms in the house had been closed off.

Now that I was "grown up," I was given a larger room—it had been Marcus's room, but he and David were now both at university. Here I had a gas fire and an old desk and bookshelves of my own, and for the first time in my life I felt I had a place, a space. I would spend hours in my room, reading, dreaming about numbers and chemistry and metals.

Above all, I delighted in being able to visit Uncle Tungsten again—his place, at least, seemed relatively unchanged (though tungsten was now in somewhat short supply, because of the vast quantities needed for making tungsten steel for armor plating). I think he also delighted in having his young protégé back, for he would spend hours with me in his factory and his lab, answering questions as fast as I could ask them. He had several glass-fronted cabinets in his office, one of which contained a series of electric lightbulbs: there were several Edison bulbs from the early 1880s, with filaments of carbonized thread; a bulb from 1897, with a fil-

---

have been unthinkable to call her by her first name; she was always Miss Levy) and always busy, she sometimes allowed me to sit by the gas fire in her little room and play while she typed my father's letters. (I loved the clack of the typewriter keys, and the little bell that rang at the end of each line.) Miss Levy lived five minutes away (in Shoot-Up Hill, a name that seemed to me more suitable perhaps for Tombstone than Kilburn), and she arrived at nine o'clock on the dot every weekday morning; she was never late, never moody or discomposed, never ill, in all the years that I knew her. Her schedule, her even presence, remained a constant through the war, even though everything else in the house had changed. She seemed impervious to the vicissitudes of life.

Miss Levy, who was a couple of years older than my father, continued to work a fifty-hour week until she was ninety, with no apparent concessions to age. Retirement was unthinkable to her, as it was to my parents, too.

ament of osmium; and several bulbs from the turn of the century, with spidery filaments of tantalum tracing a zigzag course inside them. Then there were the more recent bulbs—these were Uncle's especial pride and interest, for some of them he had pioneered himself—with tungsten filaments of all shapes and sizes. There was even one labeled "Bulb of the Future?" It had no filament, but the word *Rhenium* was inscribed on a card beside it.

I had heard of platinum, but the other metals—osmium, tantalum, rhenium—were new to me. Uncle Dave kept samples of them all, and some of their ores, in a cabinet next to the bulbs. As he handled them, he would expatiate on their unique, sovereign qualities, how they had been discovered, how they were refined, and why they were so suitable for making filaments. As Uncle spoke of the filament metals, "his" metals, they took on, in my mind, a special desirability and significance—noble, dense, infusible, glowing.

He would bring out a pitted grey nugget: "Dense, eh?" he would say, tossing it to me. "That's a platinum nugget. This is how it is found, as nuggets of pure metal. Most metals are found as compounds with other things, in ores. There are very few other metals which occur native like platinum—just gold, silver, copper, and one or two others." These other metals had been known, he said, for thousands of years, but platinum had been "discovered" only two hundred years ago, for though it had been prized by the Incas for centuries, it was unknown to the rest of the world. At first, the "heavy silver" was regarded as a nuisance, an adulterant of gold, and was dumped back into the deepest part of the river so it would not "dirty" the miners' pans again. But by the late 1700s, the new metal had enchanted all of Europe—it was denser, more ponderous than gold, and like gold it was "noble" and never tarnished. It had a luster equalling that of silver (its Spanish name, *platina,* meant "little silver").

Platinum was often found with two other metals, iridium and osmium, which were even denser, harder, more refractory. Here

Uncle pulled out samples for me to handle, mere flakes, no larger than lentils, but astoundingly heavy. They were "osmiridium," a natural alloy of osmium and iridium, the two densest substances in the world. There was something about heaviness, density—I could not say why—that gave me a thrill, and an immense sense of security and comfort. Osmium, moreover, had the highest melting point of all the platinum metals, Uncle Dave said, so at one time it was used, despite its rarity and cost, to replace the platinum filaments in lightbulbs.

The great virtue of the platinum metals was that while they were as noble and workable as gold, they had much higher melting points, and this made them ideal for chemical apparatus. Crucibles made of platinum could withstand the hottest temperatures; beakers and spatulas of it could withstand the most corrosive acids. Uncle Dave pulled out a small crucible from the cabinet, beautifully smooth and shiny. It looked new. "This was made around 1840," he said. "A century of use, and almost no wear."

My grandfather's oldest son, Jack, was fourteen years old in 1867, when diamonds were found near Kimberley in South Africa and the great diamond rush began. In the 1870s Jack, along with two brothers—Charlie and Henry (Henry was born deaf and used sign language)—went to make their lives and fortunes in South Africa as consultants in the diamond, uranium, and gold mines (their sister Rose accompanied them). In 1873 my grandfather remarried, and had thirteen more children, and the old family myths—a combination perhaps of his elder sons' stories, Rider Haggard's tales of King Solomon's mines and the old legends of the Valley of Diamonds—caused two of the next-born (Sydney and Abe) to join their half-brothers in Africa. Later still, two of the younger brothers, Dave and Mick, joined them as well, so at one point seven of the nine Landau brothers were working as mining consultants in Africa.

A photograph that hung in our house (and now hangs in mine) shows a family group taken in 1902—Grandfather, bearded and patriarchal, his second wife, Chaya, and their thirteen children. My mother appears as a little girl of six or seven, and her youngest sister, Dooggie—the youngest of the eighteen—as a ball of fluff on the ground. The images of Abe and Sydney, one can see if one looks closely, have been grafted into place (the photographer had arranged the others to make spaces for them), for they were still in South Africa at the time—detained, and perhaps endangered, by the Boer War.[2]

The elder half-brothers, married and rooted now, stayed in South Africa. They never returned to England, though tales of them constantly circulated in the family, tales heightened to the legendary by the family mythopoeia. But the younger brothers—Sydney, Abe, Mick, and Dave—returned to England when the First World War broke out, armed with exotic tales and trophies of their mining days, including minerals of all sorts.

Uncle Dave loved handling the metals and minerals in his cabinet, allowing me to handle them, expatiating on their wonders. He saw the whole earth, I think, as a gigantic natural laboratory, where heat and pressure caused not only vast geologic movements, but innumerable chemical miracles too. "Look at these diamonds," he would say, showing me a specimen from the famous Kimberley mine. "They are almost as old as the earth. They were formed thousands of millions of years ago, deep in the earth, under unimaginable pressures. Then they were brought to the surface in this kimberlite, tracking hundreds of miles from

---

[2] There were fears for all the African relatives during the Boer War, and this must have impressed my mother deeply, for more than forty years later, she would still sing or incant a little ditty from this era:

> One, two, three—relief of Kimberley
> Four, five, six—relief of Ladysmith
> Seven, eight, nine—relief of Bloemfontein

the earth's mantle, and then through the crust, very, very slowly, till they finally reached the surface. We may never see the interior of the earth directly, but this kimberlite and its diamonds are a sample of what it is like. People have tried to manufacture diamonds," he added, "but we cannot match the temperatures and pressures that are necessary."[3]

On one visit, Uncle Dave showed me a large bar of aluminum. After the dense platinum metals, I was amazed at how light it was, scarcely heavier than a piece of wood. "I'll show you something interesting," he said. He took a smaller lump of aluminum, with a smooth, shiny surface, and smeared it with mercury. All of a sudden—it was like some terrible disease—the surface broke down, and a white substance like a fungus rapidly grew out of it, until it was a quarter of an inch high, then half an inch high, and it kept growing and growing until the aluminum was completely eaten up. "You've seen iron rust—oxidizing, combining with the oxygen in the air," Uncle said. "But here, with the aluminum, it's a million times faster. That big bar is still quite shiny, because it's covered by a fine layer of oxide, and that protects it from further change. But rubbing it with mercury destroys the surface layer, so then the aluminum has no protection, and it combines with the oxygen in seconds."

I found this magical, astounding, but also a little frightening—to see a bright and shiny metal reduced so quickly to a crumbling mass of oxide. It made me think of a curse or a spell,

---

[3] There were many attempts to manufacture diamonds in the nineteenth century, the most famous being those of Henri Moissan, the French chemist who first isolated fluorine and invented the electrical furnace. Whether Moissan actually got any diamonds is doubtful—the tiny, hard crystals he took for diamond were probably silicon carbide (which is now called moissanite). The atmosphere of this early diamond-making, with its excitements, its dangers, its wild ambitions, is vividly conveyed in H. G. Wells's story "The Diamond Maker."

the sort of disintegration I sometimes saw in my dreams. It made me think of mercury as evil, as a destroyer of metals. Would it do this to every sort of metal?

"Don't worry," Uncle answered, "the metals we use here, they're perfectly safe. If I put this little bar of tungsten in the mercury, it would not be affected at all. If I put it away for a million years, it would be just as bright and shiny as it is now." The tungsten, at least, was stable in a precarious world.

"You've seen," Uncle Dave went on, "that when the surface layer is broken, the aluminum combines very rapidly with oxygen in the air to form this white oxide, which is called alumina. It is similar with iron as it rusts; rust is an iron oxide. Some metals are so avid for oxygen that they will combine with it, tarnishing, forming an oxide, the moment they are exposed to the air. Some will even pull the oxygen out of water, so one has to keep them in a sealed tube or under oil." Uncle showed me some chunks of metal with a whitish surface, in a bottle of oil. He fished out a chunk and cut it with his penknife. I was amazed at how soft it was; I had never seen a metal cut like this. The cut surface had a brilliant, silvery luster. This was calcium, Uncle said, and it was so active that it never occurred in nature as the pure metal, but only as compounds or minerals from which it had to be extracted. The white cliffs of Dover, he said, were chalk; others were made of limestone—these were different forms of calcium carbonate, a major component in the crust of the earth. The calcium metal, as we spoke, had oxidized completely, its bright surface now a dull, chalky white. "It's turning into lime," Uncle said, "calcium oxide."

But sooner or later Uncle's soliloquies and demonstrations before the cabinet all returned to *his* metal. "Tungsten," he said. "No one realized at first how perfect a metal it was. It has the highest melting point of any metal, it is tougher than steel, and it keeps its strength at high temperatures—an ideal metal!"

Uncle had a variety of tungsten bars and ingots in his office. Some he used as paperweights, but others had no discernible function whatever, except to give pleasure to their owner and maker. And indeed, by comparison, steel bars and even lead felt light and somehow porous, tenuous. "These lumps of tungsten have an extraordinary concentration of mass," he would say. "They would be deadly as weapons—far deadlier than lead."

They had tried to make tungsten cannonballs at the beginning of the century, he added, but found the metal too hard to work— though they used it sometimes for the bobs of pendulums. If one wanted to weigh the earth, Uncle Dave suggested, and to use a very dense, compact mass to "balance" against it, one could do no better than to use a huge sphere of tungsten. A ball only two feet across, he calculated, would weigh five thousand pounds.

One of tungsten's mineral ores, scheelite, Uncle Dave told me, was named after the great Swedish chemist Carl Wilhelm Scheele, who was the first to show that it contained a new element. The ore was so dense that miners called it "heavy stone" or *tung sten,* the name subsequently given to the element itself. Scheelite was found in beautiful orange crystals that fluoresced bright blue in ultraviolet light. Uncle Dave kept specimens of scheelite and other fluorescent minerals in a special cabinet in his office. The dim light of Farringdon Road on a November evening, it seemed to me, would be transformed when he turned on his Wood's lamp and the luminous chunks in the cabinet suddenly glowed orange, turquoise, crimson, green.

Though scheelite was the largest source of tungsten, the metal had first been obtained from a different mineral, called wolframite. Indeed, tungsten was sometimes called wolfram, and still retained the chemical symbol W. This thrilled me, because my own middle name was Wolf. Heavy seams of the tungsten ores were often found with tin ore, and the tungsten made it more difficult to isolate the tin. This was why, my uncle continued, they had originally called the metal wolfram—for, like a hungry animal, it "stole" the tin. I liked the name *wolfram,* its sharp, animal quality, its evocation of a ravening, mystical wolf—and thought of it as a tie between Uncle Tungsten, Uncle Wolfram, and myself, O. Wolf Sacks.

"Nature offers you copper and silver and gold native, as pure metals," Uncle would say, "and in South America and the Urals, she offers the platinum metals, too." He liked to pull out the native metals from his cabinet—twists and spangles of rosy copper; wiry, darkened silver; grains of gold panned by miners in South Africa. "Think how it must have been," he said, "seeing metal for the first time—sudden glints of reflected sunlight, sudden shinings in a rock or at the bottom of a stream!"

But most metals occurred in the form of oxides, or "earths." Earths, he said, were sometimes called calxes, and these ores were known to be insoluble, incombustible, infusible, and to be, as one eighteenth-century chemist wrote, "destitute of metallic splendour." And yet, it was realized, they were very close to metals and could indeed be converted into metals if heated with charcoal; while pure metals became calxes if heated in air. What actually occurred in these processes, however, was not understood. There can be a deep practical knowledge, Uncle said, long before theory: it was appreciated, in practical terms, how one could smelt ores and make metals, even if there was no correct understanding of what actually went on.

He would conjure up the first smelting of metal, how cavemen might have used rocks containing a copper mineral—green malachite perhaps—to surround a cooking fire and suddenly realized as the wood turned to charcoal that the green rock was bleeding, turning into a red liquid, molten copper.

We know now, he went on, that when one heats the oxides with charcoal, the carbon in the charcoal combines with their oxygen and in this way "reduces" them, leaving the pure metal. Without the ability to reduce metals from their oxides, he would say, we would never have known any metals other than the handful of native ones. There would never have been a bronze age, much less an iron age; there would never have been the fascinating discoveries of the eighteenth century, when a dozen and a half new metals (including tungsten!) were extracted from their ores.

Uncle Dave showed me some pure tungstic oxide obtained from scheelite, the same substance as Scheele and the d'Elhuyars, the discoverers of tungsten, had prepared.[4] I took the bottle from him; it contained a dense yellow powder that was surprisingly heavy, almost as heavy as iron. "All we need to do," he said, "is heat it with some carbon in a crucible until it's red-hot." He mixed the yellow oxide and the carbon together, and put the crucible in a corner of the huge furnace. A few minutes later, he withdrew it with long tongs, and as it cooled, I was able to see that an exciting change had occurred. The carbon was all gone, as was most of the yellow powder, and in their place were grains of dully shining grey metal, just as the d'Elhuyars had seen in 1783.

"There's another way we could make it," Uncle said. "It's more spectacular." He mixed the tungstic oxide with finely powdered aluminum, and then placed some sugar, some potassium perchlorate, and a little sulfuric acid on top. The sugar and perchlorate

---

[4]The d'Elhuyar brothers, Juan José and Fausto, were members of the Basque Society of Friends for Their Country, a society devoted to the cultivation of arts and sciences that would meet every evening, discussing mathematics on Monday evenings, experimenting with electrical machines and air pumps on Tuesday evenings, and so on. In 1777 the brothers were sent abroad, one to study mineralogy, the other metallurgy. Their travels took them all over Europe, and one of them, Juan José, visited Scheele in 1782.

After they returned to Spain, the brothers explored the heavy black mineral wolframite and obtained from it a dense yellow powder ("wolframic acid") which they realized to be identical to the tungstic acid Scheele had obtained from the mineral "tung-sten" in Sweden, and which, he was convinced, contained a new element. They went ahead, as Scheele had not, to heat this with charcoal, and obtained the pure new metallic element (which they named wolframium) in 1783.

and acid took fire at once, and this in turn ignited the aluminum and tungstic oxide, which burned furiously, sending up a shower of brilliant sparks. When the sparks cleared, I saw a white-hot globule of tungsten in the crucible. "That is one of the most violent reactions there is," said Uncle. "They call this the thermite process; you can see why. It can generate a temperature of three thousand degrees or more—enough to melt the tungsten. You see I had to use a special crucible lined with magnesia, to withstand the temperature. It's a tricky business, things can explode if you're not careful—and in the war, of course, they used this process to make incendiary bombs. But if conditions are right, it's a wonderful method, and it has been used to obtain all the difficult metals—chromium, molybdenum, tungsten, titanium, zirconium, vanadium, niobium, tantalum."

We scraped out the tungsten grains, washed them carefully with distilled water, examined them with a magnifying glass, and weighed them. He pulled out a tiny, 0.5-milliliter graduated cylinder, filled it to the 0.4-milliliter mark with water, then tipped in the tungsten grains. The water rose a twentieth of a milliliter. I jotted down the exact figures, and worked them out—the tungsten weighed a little less than a gram, and had a density of 19. "That's very good," Uncle said, "that's pretty much what the d'Elhuyars got when they first made it back in the 1780s.

"Now I've got several different metals here, all in little grains. Why don't you get some practice weighing these, measuring their volume, working out their density?" I spent the next hour delightedly doing this and found that Uncle had indeed given me a huge range, from one silvery metal, a little tarnished, which had a density of less than 2, to one of his osmiridium grains (I recognized it), which was almost a dozen times as dense. When I measured the density of a little yellow grain, it was exactly the same as that of tungsten—19.3, to be exact. "You see," said Uncle, "gold's density is almost the same as tungsten's, but silver is much lighter. It is easy to feel the difference between pure gold and gilded silver—but you would have a problem with gold-plated tungsten."

Scheele was one of Uncle Dave's great heroes. Not only had he discovered tungstic acid and molybdic acid (from which the new element molybdenum was made), but hydrofluoric acid, hydrogen sulfide, arsine, and prussic acid, and a dozen organic acids, too. All this, Uncle Dave said, he did by himself, with no assistants, no funds, no university position or salary, but working alone, trying to make ends meet as an apothecary in a small provincial Swedish town. He had discovered oxygen, not by a fluke, but by making it in several different ways; he had discovered chlorine; and he had pointed the way to the discovery of manganese, of barium, of a dozen other things.

Scheele, Uncle Dave would say, was wholly dedicated to his work, caring nothing for fame or money and sharing his knowledge, whatever he had, with anyone and everyone. I was impressed by Scheele's generosity, no less than his resourcefulness, by the way in which (in effect) he gave the actual discovery of elements to his students and friends—the discovery of manganese to Johan Gahn, the discovery of molybdenum to Peter Hjelm, and the discovery of tungsten itself to the d'Elhuyar brothers.

Scheele, it was said, never forgot anything if it had to do with chemistry. He never forgot the look, the feel, the smell of a substance, or the way it was transformed in chemical reactions, never forgot anything he read, or was told, about the phenomena of chemistry. He seemed indifferent, or inattentive, to most things else, being wholly dedicated to his single passion, chemistry. It was this pure and passionate absorption in phenomena—noticing everything, forgetting nothing—that constituted Scheele's special strength.

Scheele epitomized for me the romance of science. There

seemed to me an integrity, an essential goodness, about a life in science, a lifelong love affair. I had never given much thought to what I might be when I was "grown up"—growing up was hardly imaginable—but now I knew: I wanted to be a chemist. A chemist like Scheele, an eighteenth-century chemist coming fresh to the field, looking at the whole undiscovered world of natural substances and minerals, analyzing them, plumbing their secrets, finding the wonder of unknown and new metals.

5

LIGHT FOR THE MASSES

Uncle Tungsten was a complex mixture, at once intellectual and practical, as were most of his brothers and sisters, and the man who fathered them all. He loved chemistry, but he was not a "pure" chemist, like his younger brother Mick; Uncle Dave was an entrepreneur, a businessman, as well. He was a manufacturer who made a moderately good living—there was always a ready sale for his bulbs and vacuum tubes, and this was enough. He knew everyone who worked for him in friendly, personal detail. He had no desire to expand, to become huge, as he could easily have done. He remained, as he had first been, a lover of metals and materials, endlessly fascinated by their properties. He would spend hundreds of hours watching all the processes in his factories: the sintering and drawing of the tungsten, the making of the coiled coils and molybdenum supports for the filaments, the filling of the bulbs with argon in the old factory in Farringdon, and the blowing of the glass bulbs and their pearling with hydrofluoric acid at his new factory in Hoxton. He did not need to do this—his staff was competent, and the machinery worked perfectly—but he loved it, and he would sometimes think of further refinements, new processes, as he did so. He did not really need the compact but finely equipped laboratories in his factories, but he was curious and addicted to experiment, some of it with immediate application to his manufacturing, though much of it, as far as I could judge, for the pure pleasure of it, for fun. Nor did he need to know, as he did in encyclopedic detail, the history of incandescent lamps, of lighting generally, and of the basic chemistry and physics behind them. But he loved to feel that he was part of a tradition—a tradition at once of pure science, applied science, artisanry, and industry.

Edison's vision, Uncle liked to say, of light for the masses, had finally come true in the incandescent bulb. If someone could look at the earth from outer space, see how it rotated every twenty-four hours into the shadow of night, they would see millions, hundreds of millions, of incandescent bulbs light up nightly, glowing with white-hot tungsten, in the folds of that shadow—and know that man had finally conquered the darkness. The incandescent bulb had done more to alter social habits, human lives, Uncle would say, than any other invention he could think of.

And in many ways, Uncle Dave told me, the history of chemical discovery was inseparable from the quest for light. Before 1800, one had only candles or simple oil lamps such as had been used for thousands of years. Their light was feeble, and the streets were dark and dangerous, so one could hardly go out at night without a lantern or a full moon. There was a tremendous need for an efficient form of lighting that could be used safely and easily in the home and in streetlamps.

At the beginning of the nineteenth century gas lighting was introduced, and people experimented with many forms of this. Different nozzles produced gas flames of different shapes: there was the bat's-wing and the fish-tail and the cock-spur and the cock's-comb—I loved these names, as he said them, just as I loved the beautiful shapes of the flames.

But gas flames, with their glowing carbon particles, were scarcely brighter than candle flames. One needed something additional, a material that would shine with special brilliance when heated in a gas flame. Such a substance was calcia— calcium oxide, or lime—which shone with an intense greenish white light when heated. This "limelight," Uncle Dave said, was discovered in the 1820s and used to illuminate the stages in theaters for many decades—that was why we still talked about "the limelight," even though we no longer used lime for incandescence. One could get a similar brilliant light by heating several other earths—zirconia, thoria, magnesia, alumina, zinc oxide. ("Do they call that zincia?" I asked. "No," said Uncle, smiling— "I never heard it called that.")

It became clear by the 1870s, after many oxides had been tried, that there were some mixtures that glowed more brilliantly than any of the individual oxides. Auer von Welsbach, in Germany, experimented with innumerable such combinations and finally, in 1891, hit on the ideal: a 99-to-1 mixture of thoria and ceria. This ratio was critical: a 100-to-1 or 98-to-1 ratio, Auer found, was far less effective.

Up to this point, bars or pencils of oxide had been used, but Auer found that "a fabric of suitable shape," a ramie mantle, could provide a far greater surface area to be impregnated with his mixture and thus a brighter light. These mantles would revolutionize the whole industry of gas lighting, allowing it to seriously compete with the infant electric light industry.

My uncle Abe, a few years older than Uncle Dave, had a vivid memory of this discovery, and of how their somewhat dimly lit house in Leman Street was suddenly transformed by the new incandescent mantles. He remembered too how there had been a great thorium rush: in the course of a few weeks thorium rose to ten times its previous price and a desperate search began for new sources of the element.

Edison, in America, was also a pioneer experimenter on the incandescence of various rare earths, but had failed to make the breakthrough that Auer did, and had turned his attention in the late 1870s to perfecting a different sort of light, an electric light. Swan, in England, and several others, had started experimenting with platinum bulbs in the 1860s (Uncle had one of these early Swan bulbs in his cabinet); and Edison, intensely competitive, now joined the race, but found, like Swan, that there were major difficulties: platinum's melting point, though high, was not high enough.

Edison experimented with many other metals with higher melting points to get a workable filament, but none proved suitable. Then in 1879 he had a brainwave. Carbon had a much higher melting point than any metal—no one had ever been able to melt it—and though it conducted electricity, it had a high resistance, which would make it heat up and incandesce more easily. Edison tried making spirals of elemental carbon, akin to the metal spirals in earlier filaments, but these carbon spirals fell apart. His solution—almost absurdly simple, though it took an act of genius on his part to see it—was to take an organic fiber (paper, wood, bamboo, linen or cotton thread) and burn it, leaving a skeleton of carbon sufficiently strong to hold together and conduct a current. If these filaments were inserted into evacuated bulbs, they could provide a steady light for hundreds of hours.

Edison's bulbs opened up the possibility of a real revolution— though, of course, they had to be tied to a whole new system of dynamos and power lines. "The first central electrical system in the world was constructed by Edison right here in 1882," Uncle said, taking me to the window and motioning toward the streets below. "Big steam dynamos were installed on Holburn Viaduct, over there, and they supplied three thousand electric lightbulbs along the Viaduct, and on Farringdon Bridge Road."

The 1880s, then, were dominated by electric bulbs and the setting up of a whole network of power stations and power lines. But then in 1891, Auer's perfected gas mantles, which were highly efficient and moderately priced (and could use existing gas lines), mounted a serious challenge to the young industry of electric light. My uncles had told me about the fight between electric and gas lighting when they were young, and how the balance kept shifting in favor of one or the other. Many houses built in this era—including ours—had been equipped for both, as it was unclear which would win out in the end. (Even fifty years later, in my boyhood, there were many streets in London, especially in the City, that were still lit by gas mantles, and sometimes at dusk one could see the lamplighter with his tall pole, going from one streetlamp to another, lighting them one by one. I loved to watch this.)

But for all their virtues, the carbon bulbs had problems. They were fragile and became more so with use, and they could only be run at a relatively low temperature, so one had a dullish yellow light, not a brilliant white one.

Was there any way out of this? One needed a material with a melting point almost as high as carbon's, or at least around 3,000°C, but with a toughness a thread of carbon could never have—and only three such metals were known: osmium, tantalum, and tungsten. Uncle Dave seemed to become more animated at this point. He admired Edison and his ingenuity enormously, but carbon filaments, it was obvious, were not to his taste. A respectable filament, he seemed to feel, had to be made of metal, for only metals could be drawn to form proper wires. A wire of soot, he sniffed, was a contradiction in terms, and it was astounding that they held up as well as they did.

The first osmium bulbs were made in 1897 by Auer, and Uncle Dave had one of these in his cabinet. But osmium was very rare—the total world production was only fifteen pounds a year—and very costly. It was almost impossible then to draw osmium into wire, so the osmium powder had to be mixed with a binder and squirted into a mold, the binder being subsequently burned off. These osmium filaments, moreover, were very fragile and would break if the bulbs were turned upside down.

Tantalum had been known for a century or more, though there

had always been great difficulties in purifying and working it. By 1905 it became possible to purify the metal enough to draw it into wires, and with tantalum filaments, incandescent bulbs could be mass-produced cheaply and compete with carbon bulbs in a way never possible with osmium bulbs. But to get the requisite resistance, one had to use a great length of spidery-thin wire, zigzagging it inside the bulb to make a complex cagelike filament. Though tantalum softened a little when heated, these filaments were nonetheless highly successful, and finally challenged the hegemony of the gas mantle. "Suddenly," Uncle said, "tantalum bulbs were all the rage."

Tantalum bulbs continued to be the rage until the First World War, but even at the height of their popularity, another filament metal, tungsten, was being explored. The first viable tungsten lamps were made in 1911, and could operate briefly at very high temperatures, though they would soon blacken with the evaporation of tungsten and its deposition on the inner surface of the glass. This challenged the ingenuity of Irving Langmuir, an American chemist who suggested the use of an unreactive gas to exert a positive pressure on the filament, and thus reduce its evaporation. An absolutely inert gas was called for, and an obvious candidate was argon, which had been isolated fifteen years before. But the use of gas filling in turn led to another problem: massive heat loss from convection through the gas. The answer to this, Langmuir realized, was to have as compact a filament as possible, a tightly coiled helix of wire, not a spread-out spiderweb. Such a tight coil could be made with tungsten, and in 1913, all this was put together: finely drawn tungsten wire, tightly coiled helices, in a bulb filled with argon. It was evident, at this point, that the days of the tantalum bulb were numbered, and that tungsten—tougher, cheaper, more efficient—would soon replace it (although this could not happen until after the war, when argon became available in commercial quantities). It was at this point that many manufacturers turned to making tungsten bulbs, and that Uncle Dave, with several of his brothers (and three of his wife's brothers, the Wechslers, also chemists), pooled their resources and founded their firm, Tungstalite.

Uncle Dave loved telling me this saga, much of which he had lived through himself, and its pioneers were heroes to him, not least because they had been able to combine a passion for pure science with strong practical and business sense (Langmuir, he told me, was the first industrial chemist ever to get a Nobel prize).

Uncle Dave's bulbs were larger than Osram, or GE, or other electric bulbs on the market—larger, heavier, and almost absurdly robust, and they seemed to last forever. Sometimes I longed for them to expire, so I could then smash them (not easy) and pull out the tungsten filaments and their molybdenum supports, and then have the pleasure of going to the triangular cupboard under the stairs to get a new, mint bulb, wrapped in its crinkled cardboard cylinder. Other people bought their electric bulbs one at a time, but we were sent cartons straight from the factory, a few dozen bulbs at a time—60-watt and 100-watt bulbs for the most part, though we used little 15-watt bulbs for cupboards and night-lights, and a blazing 300-watt bulb as a beacon on the front porch. Uncle Tungsten made lightbulbs of all sorts and sizes, from dinky 1½-volt bulbs designed for little penlights to immense bulbs used for football fields or searchlights. There were also bulbs of special shapes, designed for instrument dials, ophthalmoscopes, and other medical instruments; and (despite Uncle's attachment to tungsten) bulbs with filaments of tantalum for use in cinema projectors and on trains. Such filaments were less efficient, less capable of higher temperatures than tungsten, but more resistant to vibration. These, too, I liked to break open when they went pfft!, so I could extract the tantalum wire inside and add it to my growing stock of metals and chemicals.

Uncle's bulbs, and my taste for improvisation, incited me to set up a lighting system of my own inside the dark cupboard under the stairs. I had always been fascinated and slightly frightened by this space, which had no light of its own and seemed to disappear, in its furthest recesses, into secrecy and mystery. I used a 6-watt bulb, lemon-shaped, of the sort used in the sidelights of our car, and a 9-volt battery designed for an electric lantern. I put a switch, rather awkwardly, on the wall and ran wires from it to the bulb and the battery. I was absurdly proud of this little installation and made a point of showing it to visitors when they came to the house. But its glare penetrated the recesses of the cupboard, and in banishing the darkness, banished its mystery, too. Too much light, I decided, was not a good thing—there were some places best left with their secrets intact.

6

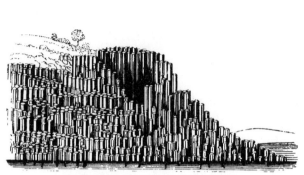

Basaltic columns, coast of Illawarra, New South Wales.

THE LAND OF STIBNITE

I think I was somewhat a loner at my new school, The Hall, at least when I first came back to London. My friend Eric Korn, who had known me before the war—we were much the same age, and would both be taken to Brondesbury Park to play by our nannies—felt that something had happened to me. I had been aggressive and normal, he said, before the war, would pick fights, stand up for myself, speak my mind; where now I seemed intimidated, timid, did not start fights or conversations, withdrew, kept my distance. I did indeed keep a distance, in almost all ways, from the school. For I was fearful of more bullying or beating, and slow to realize that school could be a good place. But I was persuaded (or forced—I can no longer remember) to join the Cub Scouts. This, it was felt, would be good for me, would make me mix with others of my own age, teach me "needed" skills for the outdoor life, like making a fire, camping, tracking—though it was not quite clear how such skills would be deployed in urban

London. And for some reason, I never really learned them. I had no sense of direction, and no visual memory—when we played Kim's Game, memorizing an array of different objects, I was so bad that there was some thought I might be mentally defective. Fires I laid could never be started, or went out within a few seconds; my attempts at making fire by rubbing two sticks together never succeeded (though I was able to conceal this, for some time, by borrowing my brother's cigarette lighter); and my attempts to pitch a tent caught universal mirth.

The only things I really liked about the Cub Scouts were the fact that we all wore the same uniform (which reduced my self-consciousness, my sense of being different), the invocations to Akela the grey wolf, and our identification with the wolf cubs in *The Jungle Book*—a gentle founding myth that pleased my romantic side. But the actual scout life, with me at least, continually miscarried in all sorts of ways.

This came to a head one day when we were asked to make a special damper like those made by Baden-Powell, the founder of the Scouts, on his sojourn in Africa. Dampers, I understood, were hard, baked discs of unleavened flour, but when I sought for flour in our kitchen I found the flour bin, as it happened, empty. I did not want to ask if there was more flour, or go out and buy some—after all, we were supposed to be resourceful and self-sufficient—so I looked around further, and then, to my pleasure, discovered some cement outside, left by builders who had been constructing a wall. I cannot now reconstruct the mental process by which I persuaded myself that cement would do instead of flour, but I used the cement, made it into a paste, flavored it (with garlic), shaped it into a damperlike oval, and baked it in the oven. It became hard, very hard—but then dampers *were* very hard. When I brought it to the Cub meeting the next day, and handed it to Mr. Baron, the scoutmaster, he was astonished, but (I think) gratified, or intrigued, by its weight, the unusually heavy nourishment it promised. He put it into his mouth and sank his teeth into it, and was rewarded with a loud cracking as one of his teeth broke. He instantly spat the thing out; there were one or two twitters, and then an awful silence: everyone in the wolf pack looked at me.

"How did you make the damper, Sacks?" Mr. Baron asked, his voice menacingly quiet. "What did you put in it?"

"I put cement, sir," I said, "I couldn't find any flour."

The silence deepened, extended; everything seemed to freeze in a sort of motionless tableau. Struggling to control himself, and (I think) not to hit me, Mr. Baron made a short, impassioned speech: I had seemed quite a nice boy, he said, decent enough, though shy, incompetent, and a terrible bungler, but this business of the damper now raised very deep questions—did I realize what I was doing, was it my intention to harm? I tried to say it was only a joke, but it was beyond me to get any words out. Was I just incredibly stupid, was I vicious, or perhaps insane? Whatever the case, I had grossly misbehaved, I had injured my master, betrayed the ideals of the wolf pack. I was not fit to be a Scout, and with this Mr. Baron summarily expelled me.

The term "acting out" had not yet been invented, but the concept was often discussed, not a mile from the school, in Anna Freud's Hampstead Clinic, where she was seeing every sort of disturbed and delinquent behavior in youngsters who had been through traumatic evacuations.

The Willesden Public Library was an odd triangular building set at an angle to Willesden Lane, a short walk from our house. It was deceptively small outside, but vast inside, with dozens of alcoves and bays full of books, more books than I had ever seen in my life. Once the librarian was assured I could handle the books and use the card index, she gave me the run of the library and allowed me to order books from the central library and even sometimes to take rare books out. My reading was voracious but unsystematic: I skimmed, I hovered, I browsed, as I wished, and though my interests were already firmly planted in the sciences, I would also, on occasion, take out adventure or detective stories as well. My school, The Hall, had no science and hence little interest for me—our curriculum, at this point, was based solely on the classics. But this did not matter, for it was my own reading in the library that provided my real education, and I divided my spare time, when I was not with Uncle Dave, between the library and the wonders of the South Kensington museums, which were crucial for me throughout my boyhood and adolescence.

The museums, especially, allowed me to wander in my own way, at leisure, going from one cabinet to another, one exhibit to another, without being forced to follow any curriculum, to attend to lessons, to take exams or compete. There was something passive, and forced upon one, about sitting in school, whereas in museums one could be active, explore, as in the world. The museums—and the zoo, and the botanical garden at Kew— made me want to go out into the world and explore for myself, be a rock hound, a plant collector, a zoologist or paleontologist. (Fifty years later, it is still natural history museums and botanical gardens I seek out whenever I visit a new city or country.)

One gained entrance to the Geological Museum, as to a temple, through a great arch of marble flanked by enormous vases of Derbyshire blue-john, a form of fluorspar. The ground floor was devoted to densely filled cabinets and cases of minerals and gems. There were dioramas of volcanoes, bubbling mudholes, lava cooling, minerals crystallizing, the slow processes of oxidation and reduction, rising and sinking, mixing, metamorphosis; so one could get not only a sense of the products of the earth's activities—its rocks, its minerals—but of the processes, physical and chemical, that continually produced them.

Up on the top floor was a colossal cluster of stibnite—glossy black, spearlike prisms of antimony sulfide. I had seen antimony sulfide as an unremarkable black powder in Uncle Dave's lab, but here I saw it in crystals five or six feet high. I worshiped these prisms; they became for me a sort of totem or fetish. These fabulous crystals, the largest of their sort in the world, had come from the Ichinokawa Mine, the legend said, on Shikoku Island, in Japan. When I grew up, I thought, when I was able to travel, I would pay a visit to this island, pay my respects to the god. Stibnite is found in many places, I subsequently learned, but that first sight joined it indissolubly with Japan in my mind, so that Japan, ever afterwards, was for me the Land of Stibnite. Australia, similarly, became the Land of Opal, no less than the Land of the Kangaroo and the Platypus.

There was a great mass of galena in the museum too—it must have weighed over a ton—which had formed in gleaming dark grey cubes five or six inches across that often had smaller cubes embedded in them. These in turn, I could see by peering through my hand lens, had yet smaller cubes seemingly growing out of them. When I mentioned this to Uncle Dave, he said that galena was cubic through and through, and that if I could look at it magnified a million times, I would still see cubes, and smaller cubes attached to these. The shape of the galena cubes, of all crystals, Uncle said, was an expression of the way their atoms were arranged, the fixed, three-dimensional patterns or lattices they formed. This was because of the bonds between them, he said, bonds that were electrostatic in nature, and the actual arrangement of atoms in a crystal lattice reflected the closest packing that the attractions and repulsions between the atoms would allow. That a crystal was built from the repetition of innumerable identical lattices—that it was, in effect, a single giant self-replicating lattice—seemed marvelous to me. Crystals were like colossal microscopes that allowed one to see the actual configuration of the atoms inside them. I could almost see, in my mind's eye, the lead atoms and the sulfur atoms composing the galena— I imagined them vibrating slightly with electrical energy, but otherwise firmly held in position, joined to one another now, coordinated in an infinite cubic lattice.

I had visions (especially after listening to stories of my uncles in their prospecting days) of being a sort of boy geologist myself, armed with chisel and hammer and collecting bags for my trophies, coming upon never-before-described mineral species. I did try a little prospecting in our garden, but found little beyond odd chips of marble and flint. I longed to go out on geological excursions, to see the patterns of the rocks, the richness of the mineral world, for myself. This desire was fanned by my reading, not only accounts of the great naturalists and explorers but also more modest books that came to hand, such as Dana's little book *The Geological Story,* with its beautiful illustrations, and my favorite nineteenth-century *Playbook of Metals,* which was subtitled *Personal Narratives of Visits to Coal, Lead, Copper and Tin Mines.* I wanted to visit different mines myself, and not just the copper and lead and tin mines in England, but the gold and diamond mines which had drawn my uncles to Africa. But failing this, the museum could provide a microcosm of the world—compact, attractive, a distillation of the experience of innumerable collectors and explorers, their material treasures, their reflections and thoughts.

I would devour the information provided in the legends for each display. Among the delights of mineralogy were the beautiful and often ancient terms used. *Vug,* Uncle Dave told me, was a term used by the old tin miners of Cornwall, and came from the Cornish dialect word *vooga* (or *fouga*), meaning an underground chamber; ultimately this came from the Latin *fovea,* a pit. It intrigued me to think that this funny, ugly word bore testament to the antiquity of mining, to the Romans' first colonization of England, drawn by the tin mines of Cornwall. The very name for tin ore, cassiterite, came from the Cassiterides, the "Tin Isles" of the Romans.

The names of minerals especially fascinated me—their sounds, their associations, the sense they gave of people and places. The older names gave one a sense of antiquity and

59

alchemy: corundum and galena, orpiment and realgar. (Orpiment and realgar, two arsenic sulfides, went euphoniously together, and made me think of an operatic couple, like Tristan and Isolde.) There was pyrites, fool's gold, in brassy, metallic cubes, and chalcedony and ruby and sapphire and spinel. Zircon sounded oriental, calomel Greek—its honeylike sweetness, its "mel," belied by its poisonness. There was the medieval-sounding sal ammoniac. There was cinnabar, the heavy red sulfide of mercury, and massicot and minium, the twin oxides of lead.

Then there were minerals named after people. One of the most common minerals, much of the redness of the world, was the hydrated iron oxide called goethite. Was this named in honor of Goethe, or did he discover it? I had read that he had a passion for mineralogy and chemistry. Many minerals were named after chemists—gay-lussite, scheelite, berzelianite, bunsenite, liebigite, crookesite, and the beautiful, prismatic "ruby-silver," proustite. There was samarskite, named after a mining engineer, Colonel Samarski. There were other names that were evocative in a more topical way: stolzite, a lead tungstate, and scholzite, too. Who were Stolz and Scholz? Their names seemed very Prussian to me, and this, just after the war, evoked an anti-German feeling. I imagined Stolz and Scholz as Nazi officers with barking voices, sword sticks, and monocles.

Other names appealed to me simply for their sound or for the images they conjured up. I loved classical words and their depiction of simple properties—the crystal forms, colors, shapes, and optics of minerals—like diaspore and anastase and microlite and polycrase. A great favorite was cryolite—ice stone, from Greenland, so low in refractive index that it was transparent, almost ghostly, and, like ice, became invisible in water.[1]

---

[1] Cryolite was the chief mineral in a vast pegmatitic mass in Ivigtut, Greenland, and this ore was mined continuously for more than a century. The miners, who had sailed from Denmark, would sometimes take boulders of the trans-

Many elements had been given names from folklore or mythology, sometimes revealing a little of their history. A *kobold* was a goblin or evil sprite, a *nickel* a devil; both were terms used by Saxon miners when cobalt and nickel ores proved treacherous, and did not yield what they should. Tantalum brought up visions of Tantalus tantalized in Hell by water that retreated from him whenever he bent down to drink from it; the element was given its name, I read, because its oxide was unable to "drink water," that is, to dissolve in acids. Niobium was named after Tantalus' daughter, Niobe, because the two elements were always found together. (My 1860ish books included a third element, pelopium, in this family—Pelops was Tantalus' son, whom he cooked and served up to the gods—but the existence of this was later disproved.)

Other elements had astronomical names. There was uranium, discovered in the eighteenth century and named after the planet Uranus; and a few years later, palladium and cerium, named after the recently discovered asteroids Pallas and Ceres. Tellurium had a fine, earthy Greek name, and it was only natural that when its lighter analog was found, it should be named selenium, after the moon.[2]

---

parent cryolite to use as anchors for their boats, and never quite got used to the way in which these vanished, became invisible, the instant they sank below the surface of the water.

[2] In addition to the hundred-odd names of existing elements, there were at least twice that number for elements that never made it, elements imagined or claimed to exist on the basis of unique chemical or spectroscopic characteristics, but later found to be known elements or mixtures. Many were place names, often exotic, discarded because the elements turned out to be spurious: "florentium," "moldavium," "norwegium," and "helvetium," "austrium" and "russium," "illinium," "virginium," and "alabamine," and the splendidly named "bohemium."

I was oddly moved by these fictional elements and their names, especially the starry ones. The most beautiful, to my ears, were "aldebaranium" and "cassiopeium" (Auer's names for elements that actually existed, ytterbium and

I loved to read of the elements and their discovery—not just the chemical, but the human aspects of this enterprise, and all this, and more, I learned from a delightful book published just before the war by Mary Elvira Weeks, *The Discovery of the Elements.* Here I got a vivid idea of the lives of many chemists, the great variety, and sometimes vagaries, of character they showed. And here I found quotations from the early chemists' letters, which portrayed their excitements and despairs as they fumbled and groped their way to their discoveries, losing the track now and again, getting caught in blind alleys, though ultimately reaching the goals they sought.

My history and geography as a boy, the history and geography that moved me, was based more on chemistry than on wars or world events. I followed the fortunes of the early chemists more closely than the fortunes of the contending forces in the war (perhaps, indeed, they helped insulate me from the frightening realities around me). I longed to go to "ultima Thule," the far-northern home of the element thulium, and to visit the little village of Ytterby in Sweden, which had given its name to no fewer than four elements (ytterbium, terbium, erbium, yttrium). I longed to go to Greenland, where, I imagined, there were whole mountain ranges, transparent, scarcely visible, of ghostly cryo-

---

lutecium) and "denebium," for a mythical rare earth. There had been a "cosmium" and "neutronium" ("element 0"), too, to say nothing of "archonium," "asterium," "aetherium," and the Ur-element "anodium," from which all the other elements supposedly were built.

There were sometimes competing names for new discoveries. Andrés del Rio discovered vanadium in Mexico in 1800 and named it "panchromium" for the variety of its many-colored salts. But other chemists doubted his discovery, and he eventually gave up his claim, and the element was only rediscovered and renamed thirty years later by a Swedish chemist, this time in honor of Vanadis, the Norse goddess of beauty. Other obsolete or discredited names also referred to actual elements: thus the magnificent "jargonium," an element supposedly present in zircons and zirconium ores, was most probably the real element hafnium.

lite. I longed to go to Strontian, in Scotland, to see the little village that had given strontium its name. The whole of Britain, for me, could be seen in terms of its many lead minerals—there was matlockite, named for Matlock in Derbyshire; leadhillite, named after the Leadhills in Lanarkshire; lanarkite, also from Lanarkshire; and the beautiful lead sulfate, anglesite, from Anglesey in Wales. (There was also the town of Lead in South Dakota—a town, I liked to imagine, actually built of metallic lead.) The geographic names of elements and minerals stood out for me like lights over the map of the world.

Seeing the minerals in the museum incited me to get little bags of "mixed minerals" from a local shop for a few pennies; these would contain little pieces of pyrites, galena, fluorite, cuprite, hematite, gypsum, siderite, malachite, and different forms of quartz, to which Uncle Dave might contribute rarer things, like tiny fragments of scheelite which had broken off his larger piece. Most of my mineral specimens were rather battered, often tiny ones that a real collector would sniff at, but they gave me a feeling of having a sample of nature for myself.

It was from looking at minerals in the Geological Museum and studying their chemical formulas that I learned about their composition. Some were simple and invariable in composition— this was true of cinnabar, a mercury sulfide that always contained the same proportion of mercury and sulfur, no matter where a particular specimen was found. But it was different with many other minerals, including Uncle Dave's favorite scheelite. While scheelite was ideally pure calcium tungstate, some specimens contained a certain amount of calcium molybdate as well. Pure calcium molybdate, conversely, occurred naturally as the mineral powellite, but some specimens of powellite also contained small amounts of calcium tungstate. One might, in fact, have any intermediate between the two, from a mineral that was 99 percent tungstate and 1 percent molybdate to one that was 99 percent molybdate and 1 percent tungstate. This was because tungsten and molybdenum had atoms, ions, of similar size, so that an ion of one element could replace the other within the mineral's crystal lattice. But above all, it was because tungsten and molybdenum belonged to the same chemical group or family, and nature treated them, with their similar chemical and physical properties, very much alike. Thus both tungsten and molybdenum tended to form similar compounds with other elements, and both tended to occur naturally as acidic salts that crystallized from solution under similar conditions.

These two elements formed a natural pair, were chemical brothers. This fraternal relationship was even closer with the elements niobium and tantalum, which usually occurred together in the same minerals. And the fraternity approached identical twinship in the elements zirconium and hafnium, which not only invariably occurred together in the same minerals, but were so similar chemically that it took a century to distinguish them— Nature herself could hardly do so.

Wandering through the Geological Museum, I also got a sense of the enormous range, the thousands of different minerals in the earth's crust, and of the relative abundances of the elements that made them up. Oxygen and silicon were overwhelmingly common—there were more silicate minerals than any others, to say nothing of all the world's sands. And with the standard rocks of the world—the chalks and feldspars, granites and dolomites— one could see that magnesium, aluminum, calcium, sodium, and potassium must make up nine-tenths or more of the earth's crust. Iron, too, was common; there seemed to be whole areas of Australia as iron-red as Mars. And I could add little fragments of all these elements, in the form of minerals, to my own collection.

The eighteenth century, Uncle told me, had been a grand time for the discovery and isolation of new metals (not only tungsten, but a dozen others, too), and the greatest challenge to eighteenth-century chemists was how to separate these new metals from their ores. This is how chemistry, real chemistry, got on its feet, investigating countless different minerals, analyzing them, breaking them down, to see what they contained. Real chemical analysis—seeing what minerals would react with, or how they behaved when heated or dissolved—of course required a laboratory, but there were elementary observations one could do almost anywhere. One could weigh a mineral in one's hand, estimate its density, observe its luster, the color of its streak on a porcelain plate. Hardness varied hugely, and one could easily get a rough approximation—talc and gypsum one could scratch with a fingernail; calcite with a coin; fluorite and apatite with a steel knife; and orthoclase with a steel file. Quartz would scratch glass, and corundum would scratch anything but diamond.

A classical way of determining the relative density or specific gravity of a specimen was to weigh a fragment of mineral twice, in air and in water, to give the ratio of its density to that of water. Another, simpler way, and one which gave me a peculiar pleasure, was to examine the buoyancy of different minerals in liquids of different specific gravity—"heavy" liquids had to be used here, for all minerals, except ice, were denser than water. I got a series of heavy liquids: first bromoform, which was almost three times as dense as water, then methylene iodide, which was even denser, and then a saturated solution of two thallium salts called Clerici solution. This had a specific gravity of well over four, and even though it looked like ordinary water, many minerals and even some metals would easily float in it. I loved taking my little bottle of Clerici solution to school, asking people to hold it, and seeing their look of surprise as they experienced its weight, almost five times what they might have expected.

I was on the shy side at school (one school report called me "diffident"), and Braefield had added a special timidity, but when I had a natural wonder—whether it was shrapnel from a bomb;

or a piece of bismuth with its terraces of prisms resembling a miniature Aztec village; or my little bottle of arm-droppingly dense, sensorily stunning, Clerici solution; or gallium, which melted in the hand (I later got a mold, and made a teaspoon of gallium, which would shrink and melt as one stirred the tea with it)—I lost all my diffidence, and freely approached others, all my fear forgotten.

7

CHEMICAL RECREATIONS

My parents and my brothers had introduced me, even before the war, to some kitchen chemistry: pouring vinegar on a piece of chalk in a tumbler and watching it fizz, then pouring the heavy gas this produced, like an invisible cataract, over a candle flame, putting it out straightaway. Or taking red cabbage, pickled with vinegar, and adding household ammonia to neutralize it. This would lead to an amazing transformation, the juice going through all sorts of colors, from red to various shades of purple, to turquoise and blue, and finally to green.

After the war, with my new interest in minerals and colors, my brother David, who had grown some crystals when he did chemistry at school, showed me how to do this myself. He showed me how to make a supersaturated solution by dissolving a salt like alum or copper sulfate in very hot water and then letting it cool.

One needed to hang something—a thread or a bit of metal—in the solution to start the process off. I did this first with a thread of wool in a copper sulfate solution, and in a few hours this produced a beautiful chain of bright blue crystals climbing along the thread. But if I used an alum solution and a good seed crystal to start it off, I discovered, the crystal would grow evenly, on every face, giving me a single large, perfectly octahedral crystal of alum.

I later commandeered the kitchen table to make a "chemical garden," sowing a syrupy solution of sodium silicate, or water-glass, with differently colored salts of iron and copper and chromium and manganese. This produced not crystals but twisted, plantlike growths in the water-glass, distending, budding, bursting, continually reshaping themselves before my eyes.[1] This sort of growth, David told me, was due to osmosis, the gelatinous silica of the water-glass acting as a "semipermeable membrane," allowing water to be drawn in to the concentrated mineral solution inside it. Such processes, he said, were crucial in living organisms, though they occurred in the earth's

---

[1] Thomas Mann provides a lovely description of silica gardens in *Doctor Faustus:*

> I shall never forget the sight. The vessel . . . was three-quarters full of slightly muddy water—that is, dilute water-glass—and from the sandy bottom there strove upwards a grotesque little landscape of variously coloured growths: a confused vegetation of blue, green, and brown shoots which reminded one of algae, mushrooms, attached polyps, also moss, then mussels, fruit pods, little trees or twigs from trees, here and there of limbs. It was the most remarkable sight I ever saw, and remarkable not so much for its appearance, strange and amazing though that was, as on account of its profoundly melancholy nature. For when Father Leverkühn asked us what we thought of it and we timidly answered him that they might be plants: "No," he replied, "they are not, they only act that way. But do not think the less of them. Precisely because they do, because they try to as hard as they can, they are worthy of all respect."

crust as well, and this reminded me of the gigantic nodular, kidneylike masses of hematite I had seen in the museum—the label said this was "kidney ore" (though Marcus had once told me they were the fossilized kidneys of dinosaurs).

I enjoyed these experiments, and tried to envisage the processes that were going on, but I did not feel a real chemical passion—a desire to compound, to isolate, to decompose, to see substances changing, familiar ones disappearing and new ones in their stead—until I saw Uncle Dave's lab and his passion for experiments of all kinds. Now I longed to have a lab of my own—not Uncle Dave's bench, not the family kitchen, but a place where I could do chemical experiments undisturbed, by myself. As a start, I wanted to lay hands on cobaltite and niccolite, and compounds or minerals of manganese and molybdenum, of uranium and chromium—all those wonderful elements which were discovered in the eighteenth century. I wanted to pulverize them, treat them with acid, roast them, reduce them—whatever was necessary—so I could extract their metals myself. I knew, from looking through a chemical catalog at the factory, that one could buy these metals already purified, but it would be far more fun, far more exciting, I reckoned, to make them myself. This way, I would enter chemistry, start to discover it for myself, in much the same way as its first practitioners did—I would live the history of chemistry in myself.

And so I set up a little lab of my own at home. There was an unused back room I took over, originally a laundry room, which had running water and a sink and drain and various cupboards and shelves. Conveniently, this room led out to the garden, so that if I concocted something that caught fire, or boiled over, or emitted noxious fumes, I could rush outside with it and fling it on the lawn. The lawn soon developed charred and discolored patches, but this, my parents felt, was a small price to pay for my safety—their own, too, perhaps. But seeing occasional flaming globules rushing through the air, and the general turbulence and abandon with which I did things, they were alarmed, and urged me to plan experiments and to be prepared to deal with fires and explosions.

Uncle Dave advised me closely on the choice of apparatus — test tubes, flasks, graduated cylinders, funnels, pipettes, a Bunsen burner, crucibles, watch glasses, a platinum loop, a desiccator, a blowpipe, a retort, a range of spatulas, a balance. He advised me too on basic reagents — acids and alkalis, some of which he gave me from his own lab, along with a supply of stoppered bottles of all sizes, bottles of varied shapes and colors (dark green or brown for light-sensitive chemicals), with perfectly fitting ground-glass stoppers.

Every month or so, I stocked my lab with visits to a chemical supply house far out in Finchley, housed in a large shed set at a distance from its neighbors (who viewed it, I imagined, with a certain trepidation, as a place that might explode or exhale poisonous fumes at any moment). I would hoard my pocket money for weeks — occasionally one of my uncles, approving my secret passion, would slip me a half crown or so — and then take a succession of trains and buses to the shop.

I loved to browse through Griffin & Tatlock as one would browse through a bookshop. The cheaper chemicals were kept in huge stoppered urns of glass; the rarer, more costly substances were kept in smaller bottles behind the counter. Hydrofluoric acid — dangerous stuff, used for etching glass — could not be kept in glass, so it was sold in special small bottles made of gummy brown gutta-percha. Beneath the serried urns and bottles on the shelves were great carboys of acid — sulfuric, nitric, aqua regia; globular china bottles of mercury (seven pounds of this would fit into a bottle the size of a fist), and slabs and ingots of the commoner metals. The shopkeepers soon got to know me — an intense and rather undersized schoolboy, clutching his pocket money, spending hours amid the jars and bottles — and though they would warn me now and then, "Go easy with that one!" they always let me have what I wished.

My first taste was for the spectacular—the frothings, the incandescences, the stinks and the bangs, which almost define a first entry into chemistry. One of my guides was J. J. Griffin's *Chemical Recreations,* an 1850ish book I had found in a second-hand bookshop. Griffin had an easy, practical, and above all playful style; chemistry was clearly fun for him, and he made it fun for his readers, readers who must often have been, I decided, boys like myself, for he had sections like "Chemistry for the Holidays"—this included the "Volatile Plum Pudding" ("when the cover is removed . . . it leaves its dish and rises to the ceiling"), "A Fountain of Fire" (using phosphorus—"the operator must take care not to burn himself"), and "Brilliant Deflagration" (here, too, one was warned to "remove your hand instantly"). I was amused by the mention of a special formula (sodium tungstate) to render ladies' dresses and curtains incombustible—were fires that common in Victorian times?—and used it to fireproof a handkerchief for myself.

The book opened with "Elementary Experiments," experiments first with vegetable dyes, seeing their color changes with acids and alkalis. The most common vegetable dye was litmus—it came from a lichen, Griffin said. I used some of the litmus papers that my father kept in his dispensary, and saw how they turned red with different acids or blue with alkaline ammonia.

Griffin suggested experiments with bleaching—here I used my mother's bleaching powder in place of the chlorine water he suggested, and with this I bleached litmus paper, cabbage juice, and a red handkerchief of my father's. Griffin also suggested holding a red rose over burning sulfur, so that the sulfur dioxide produced would bleach it. Dipping it into water, miraculously, restored its color.

From here Griffin moved (and I with him) to "sympathetic inks," which became visible only when heated or specially treated. I played with a number of these—lead salts, which turned black with hydrogen sulfide; silver salts, which blackened when exposed to light; cobalt salts, which became visible when dried or heated. All this was fun, but it was chemistry, too.

There were other old chemistry books lying around the house, some of which had been my parents' when they were medical students, and some, more recent, belonging to my older brothers Marcus and David. One such was Valentin's *Practical Chemistry,* a workhorse of a book—straight, uninspired, pedestrian in tone, designed as a practical manual, but nevertheless, for me, filled with wonders. Inside its cover, corroded, discolored, and stained (for it had done time in the lab in its day), it bore the words "Best wishes and congratulations 21/1/13—Mick"—it had been given to my mother on her eighteenth birthday by her twenty-five-year-old brother Mick, already a research chemist himself. Uncle Mick, a younger brother of Dave, had gone to South Africa with his brothers, and then worked in a tin mine on his return. He loved tin, I was told, as much as Uncle Dave loved tungsten, and he was sometimes referred to in the family as Uncle Tin. I never knew Uncle Mick, for he died of a malignancy the year I was born—he was only forty-five—a victim, his family thought, of the high levels of radioactivity in the uranium mines in Africa. But my mother had been very close to him, and his memory and image stayed vividly in her mind. The notion that this was my mother's own chemistry book, and of the never-known, young chemist uncle who gave it to her, made the book especially precious to me.

There was a great popular interest in chemistry in the Victorian era, and many households had their own labs, as they had their ferneries and stereoscopes. Griffin's *Chemical Recreations* had originally been published around 1830 and was so popular that it was continually revised and brought out in new editions; I had the tenth, published in 1860.[2]

A companion volume to Griffin's, published at much the same time and in the same green and gilt binding, was *The Science of Home Life,* by A. J. Bernays, which focused on coal, coal gas, candles, soap, glass, china, earthenware, disinfectants—everything that might be contained in a Victorian home (and much of which was still contained in houses a century later).

Very different in style and content, though equally designed to awake the sense of wonder ("The common life of man is full of Wonders, Chemical and Physiological. Most of us pass through this life without seeing or being sensible of them . . .") was *The Chemistry of Common Life,* by J.F.W. Johnston, written in 1859. This had fascinating chapters on "The Odours We Enjoy," "The Smells We Dislike," "The Colours We Admire," "The Body We Cherish," "The Plants We Rear," and no less than eight chapters on "The Narcotics We Indulge In." This introduced me not only to chemistry, but to a panorama of exotic human behaviors and cultures.

A much earlier book, of which I was able to get a battered copy for sixpence—it had no covers, and a few pages missing—was *The Chemical Pocket-Book or Memoranda Chemica,* written in 1803. The author was a James Parkinson, of Hoxton, whom I would reencounter in my biology days as the founder of paleontology, and then again, when I was a medical student, as the author of the famous *Essay on the Shaking Palsy*—which came to be known as Parkinson's disease. But for me, at eleven, he was just the

---

[2] Griffin was not only an educator at many levels—he wrote *The Radical Theory in Chemistry* and *A System of Crystallography,* both more technical than his *Recreations*—but also a manufacturer and purveyor of chemical apparatus: his "chemical and philosophical apparatus" was used throughout Europe. His firm, later to become Griffin & Tatlock, was still a major supplier a century later, when I was a boy.

author of this delightful little pocket book of chemistry. I got a strong sense, from his book, of how chemistry was expanding, almost explosively, at the beginning of the nineteenth century; thus Parkinson spoke of ten new metals—uranium, tellurium, chromium, columbium (niobium), tantalum, cerium, palladium, rhodium, osmium, iridium—all having been discovered in the preceding few years.

It was from Griffin that I first gained a clear idea of what was meant by "acids" and "alkalis" and how they combined to produce "salts." Uncle Dave demonstrated the opposition of acids and bases by measuring out precise quantities of hydrochloric acid and caustic soda, which he mixed in a beaker. The mixture became extremely hot, but when it cooled, he said, "Now try it, drink it." Drink it—was he mad? But I did so, and tasted nothing but salt. "You see," he explained, "an acid and a base come together, and they neutralize each other; they combine and make a salt."

Could this miracle happen in reverse, I asked? Could salty water be made to produce the acid and the base all over again? "No," Uncle said, "that would require too much energy. You saw how hot it got when the acid and base reacted—the same amount of heat would be needed to reverse the reaction. And salt," he added, "is very stable. The sodium and chloride hold each other tightly, and no ordinary chemical process will break them apart. To break them apart you have to use an electric current."

He showed me this more dramatically one day by putting a piece of sodium in a jar full of chlorine. There was a violent conflagration, the sodium caught fire and burned, weirdly, in the yellowish green chlorine—but when it was over, the result was nothing more than common salt. I had a heightened respect for salt, I think, after that, having seen the violent opposites that came together in its making and the strength of the energies, the elemental forces, that were now locked in the compound.

Here, too, Uncle Dave showed me, the proportions had to be exact: 23 parts of sodium, by weight, to 35.5 of chlorine. I was struck by these numbers, for they were already familiar: I had seen them in lists in my books; they were the "atomic weights" of these elements. I had learned these numbers by rote, in the same mindless way one learns multiplication tables. But when Uncle Dave brought up these selfsame numbers in relation to the chemical combination of two elements, a slow, underground questioning started in my head.

In addition to my collection of mineral samples, I had a collection of coins, housed in a small wooden cabinet of highly polished mahogany, with doors that opened like the doors of a toy theater, revealing a series of slim trays with velvet-covered circles for the coins—some as small as a quarter-inch across (this for groats, for silver threepenny pieces, and for Maundy money, tiny silver coins given on Easter to the poor), others almost two inches across (for crowns, which I loved, and even larger than these, the gigantic twopenny pieces made at the end of the eighteenth century).

There were also stamp albums, and the stamps I most loved were those of remote islands, with pictures of local scenes and plants, stamps which could themselves provide a vicarious voyage. I adored stamps showing different minerals, and peculiar stamps of various sorts—triangular ones, imperforate ones, stamps with inverted watermarks or missing letters or advertisements printed on the back. One of my favorites was a strange Serbo-Croat stamp from 1914 which was said to show the features of the murdered Archduke Ferdinand when viewed from a certain angle.

But the collection closest to my heart was a singular collection of bus tickets. Whenever one took a bus in London in those days, one got a colored oblong of cardboard bearing letters and numbers. It was after getting an O 16 and an S 32 (my initials, also the symbols of oxygen and sulfur—and added to these, by a happy chance, their atomic weights, too) that I decided to make a collection of "chemical" bus tickets, to see how many of the ninety-two elements I could get. I was extraordinarily lucky, so it seemed to me (though there was nothing but chance involved), for the tickets accrued rapidly, and I soon had a whole collection (W 184, tungsten, gave me particular pleasure, partly because it provided my missing middle initial). There were, to be sure, some difficult ones: chlorine, irritatingly, had an atomic weight of 35.5, which was not a whole number, but, undismayed, I collected a Cl 355 and inked in a tiny decimal point. The single letters were easier to get—I soon had an H 1, a B 11, a C 12, an N 14, and an F 19, besides the original O 16. When I realized that atomic numbers were even more important than atomic weights, I started to collect these as well. Eventually, I had all the known elements, from H 1 to U 92. Every element became indissolubly associated with a number for me, and every number with an element. I loved carrying my little collection of chemical bus tickets with me; it gave me the sense that I had, in the space of a single cubic inch, the whole universe, its building blocks, in my pocket.

8

STINKS AND BANGS

ttracted by the sounds and flashes and smells coming from my lab, David and Marcus, now medical students, sometimes joined me in experiments—the nine- and ten-year age differences between us hardly mattered at these times. On one occasion, as I was experimenting with hydrogen and oxygen, there was a loud explosion, and an almost invisible sheet of flame, which blew off Marcus's eyebrows completely. But Marcus took this in good part, and he and David often suggested other experiments.

We mixed potassium perchlorate with sugar, put it on the back step, and banged it with a hammer. This caused a most satisfying explosion. It was trickier with nitrogen tri-iodide, easily made by adding concentrated ammonia to iodine, catching the nitrogen tri-iodide on filter paper, and drying it with ether. Nitrogen tri-iodide was incredibly touch-sensitive; one had only to touch it with a stick—a *long* stick (or even a feather)—and it would explode with surprising violence.

We made a "volcano" together with ammonium dichromate, setting fire to a pyramid of the orange crystals, which then flamed, furiously, becoming red-hot, throwing off showers of sparks in all directions, and swelling portentously, like a miniature volcano erupting. Finally, when it had died down, there was, in place of the neat pyramid of crystals, a huge fluffy pile of dark green chromic oxide.

Another experiment, suggested by David, involved pouring concentrated, oily sulfuric acid on a little sugar, which instantly turned black, heated, steamed, and expanded, forming a monstrous pillar of carbon rising high above the rim of the beaker. "Beware," David said, as I gazed at this transformation. "*You'll* be turned into a pillar of carbon if you get the acid on yourself." And then he told me horror stories, probably invented, of vitriol throwings in East London, and patients he had seen coming into the hospital with their entire faces all but burned off. (I was not quite sure whether to believe him, for when I was younger he had told me that if I looked at the Kohanim as they were blessing us in the shul—their heads were covered with a large shawl, a tallis, as they prayed, for they were irradiated, at this moment, by the blinding light of God—my eyes would melt in their sockets and run down my cheeks like fried eggs.)[1]

I spent a good deal of my time in the lab examining chemical colors and playing with them. There were certain colors that held

---

[1] I read John Hersey's *Hiroshima* a few years later, and I was struck by this passage:

> When he had penetrated the bushes, he saw there were about twenty men, and they were all in exactly the same nightmarish state: their faces were wholly burned, their eyesockets were hollow, the fluid from their melted eyes had run down their cheeks. (They must have had their faces upturned when the bomb went off. . . .)

a special, mysterious power for me—this was especially so of very deep and pure blues. As a child I had loved the strong, bright blue of the Fehling's solution in my father's dispensary, just as I had loved the cone of pure blue at the center of a candle flame. I found I could produce very intense blues with some cobalt compounds, with cuprammonium compounds, and with complex iron compounds like Prussian blue.

But the most mysterious and beautiful of all the blues for me was that produced by dissolving alkali metals in liquid ammonia (Uncle Dave showed me this). The fact that metals *could* be dissolved at all was startling at first, but the alkali metals were all soluble in liquid ammonia (some to an astounding degree—cesium would completely dissolve in a third its weight of ammonia). When the solutions became more concentrated, they suddenly changed character, turning into lustrous bronze-colored liquids that floated on the blue—and in this state they conducted electricity as well as a liquid metal like mercury. The alkaline earth metals would work as well, and it did not matter whether the solute was sodium or potassium, calcium or barium—the ammoniacal solutions, in every case, were an identical deep blue, suggesting the presence of some substance, some structure, something common to them all. It was like the color of the azurite in the Geological Museum, the very color of heaven.

Many of the so-called transition elements infused their compounds with characteristic colors—most cobalt and manganese salts were pink; most copper salts deep blue or greenish blue; most iron salts pale green and nickel salts a deeper green. Similarly, in minute amounts, transition elements gave many gems their particular colors. Sapphires, chemically, were basically nothing but corundum, a colorless aluminum oxide, but they could take on every color in the spectrum—with a little bit of chromium replacing some of the aluminum, they would turn ruby red; with a little titanium, a deep blue; with ferrous iron, green; with ferric iron, yellow. And with a little vanadium, the corundum began to resemble alexandrite, alternating magically between red and green—red in incandescent light, green in daylight. With certain elements, at least, the merest smattering of atoms could produce a characteristic color. No chemist could have "flavored" corundum with such delicacy, a few atoms of this, a few ions of that, to produce an entire spectrum of colors.

There were only a handful of these "coloring" elements— titanium, vanadium, chromium, manganese, iron, cobalt, nickel, and copper, so far as I could see, being the main ones. They were, I could not help noticing, all bunched together in terms of atomic weight—though whether this meant anything, or was just a coincidence, I had no idea at the time. It was characteristic of all of these, I learned, that they had a number of possible valency states, unlike most of the other elements, which had only one. Sodium, for instance, would combine with chlorine in only one way, one atom of sodium to one of chlorine. But there were two combinations of iron and chlorine: an atom of iron could combine with two atoms of chlorine to form ferrous chloride ($FeCl_2$) or with three atoms of chlorine to form ferric chloride ($FeCl_3$). These two chlorides were very different in many ways, including color.

Because it had four strikingly different valencies or oxidation states, and it was easy to transform these into one another, vanadium was an ideal element to experiment with. The simplest way of reducing vanadium was to start with a test tube full of (pentavalent) ammonium vanadate in solution and add small lumps of zinc amalgam. The amalgam would immediately react, and the solution would turn from yellow to royal blue (the color of tetravalent vanadium). One could remove the amalgam at this point, or let it react further, till the solution turned green, the color of trivalent vanadium. If one waited still longer, the green would disappear and be replaced by a beautiful lilac, the color of divalent vanadium. The reverse experiment was even more beautiful, especially if one layered potassium permanganate, a deep purple layer, over the delicate lilac; this would be oxidized over a period of hours and form separate layers, one above the other, of lilac divalent vanadium on the bottom, then green trivalent vanadium, then blue tetravalent vanadium, then yellow pentavalent vanadium (and on top of this, a rich brown layer of the original permanganate, now brown because it was mixed with manganese dioxide).

These experiences with color convinced me that there was a very intimate (if unintelligible) relation between the atomic character of many elements and the color of their compounds or minerals. The same color would show itself whatever compound one looked at. It could be, for example, manganous carbonate, or nitrate, or sulfate, or whatever—all had the identical pink of the divalent manganous ion (the permanganates, by contrast, where the manganese ion was heptavalent, were all deep purple). And from this I got a vague feeling—it was certainly not one that I could formulate with any precision at the time—that the color of these metal ions, their chemical color, was related to the specific state of their atoms as they moved from one oxidation state to another. What was it about the transition elements, in particular, that gave them their characteristic colors? Were these substances, their atoms, in some way "tuned"?[2]

A lot of chemistry seemed to be about heat—sometimes a demand for heat, sometimes the production of heat. Often one needed heat to start a reaction, but then it would go by itself,

---

[2] Such thoughts about "tuning," I was later to read, had first been raised in the eighteenth century by the mathematician Euler, who had ascribed the color of objects to their having "little particles" on their surface—atoms—tuned to respond to light of different frequencies. Thus an object would look red because its "particles" were tuned to vibrate, resonate, to the red rays in the light that fell on it:

> The nature of the radiation by which we see an opaque object does not depend on the source of light but on the vibratory motion of the very

sometimes with a vengeance. If one simply mixed iron filings and sulfur, nothing happened—one could still pull out the iron filings from the mixture with a magnet. But if one started to heat the mixture, it suddenly glowed, became incandescent, and something totally new—iron sulfide—was created. This seemed a basic, almost primordial reaction, and I imagined that it occurred on a vast scale in the earth, where molten iron and sulfur came into contact.

One of my earliest memories (I was only two at the time) was of seeing the Crystal Palace burn. My brothers took me to see it from Parliament Hill, the highest point on Hampstead Heath, and all around the burning palace the night sky was lit up in a wild and beautiful way. And every November 5, in memory of Guy Fawkes, we would have fireworks in the garden—little sparklers full of iron dust; Bengal lights in red and green; and bangers, which made me whimper with fear and want to crawl, as our dog would, under the nearest shelter. Whether it was these experiences, or whether it was a primordial love of fire, it was flames and burnings, explosions and colors, which had such a special (and sometimes fearful) attraction for me.

---

small particles [atoms] of the object's surface. These little particles are like stretched strings, tuned to a certain frequency, which vibrate in response to a similar vibration of the air even if no one plucks them. Just as the stretched string is excited by the same sound that it emits, the particles of the surface begin to vibrate in tune with the incident radiation and to emit their own waves in every direction.

David Park, in *The Fire Within the Eye: A Historical Essay on the Nature and Meaning of Light*, writes of Euler's theory:

I think this was the first time anyone who believed in atoms ever suggested that they have a vibrating internal structure. The atoms of Newton and Boyle are clusters of hard little balls, Euler's atoms are like musical instruments. His clairvoyant insight was rediscovered much later, and when it was, nobody remembered who had it first.

I liked mixing iodine and zinc, or iodine and antimony—no added heat was needed here—and seeing how they heated up spontaneously, sending a cloud of purple iodine vapor above them. The reaction was more violent if one used aluminum rather than zinc or antimony. If I added two or three drops of water to the mixture, it would catch fire and burn with a violet flame, spreading fine brown iodide powder over everything.

Magnesium, like aluminum, was a metal whose paradoxes intrigued me: strong and stable enough in its massive form to be used in airplane and bridge construction, but almost terrifyingly active once oxidation, combustion, got started. One could put magnesium in cold water, and nothing would happen. If one put it in hot water, it would start to bubble hydrogen; but if one lit a length of magnesium ribbon, it would continue to burn with dazzling brilliance *under* the water, or even in normally flame-suffocating carbon dioxide. This reminded me of the incendiary bombs used during the war, and how they could not be quenched by carbon dioxide or water, or even by sand. Indeed, if one heated magnesium with sand, silicon dioxide—and what could be more inert than sand?—the magnesium would burn brilliantly, pulling the oxygen out of the sand, producing elemental silicon or a mixture of silicon with magnesium silicide. (Nonetheless, sand was used to suffocate ordinary fires that had been started by incendiary bombs, even if it was useless against burning magnesium itself, and one saw sand buckets everywhere in London during the war; every house had its own.) If one then tipped the silicide into dilute hydrochloric acid, it would react to form a spontaneously inflammable gas, hydrogen silicide, or silane—bubbles of this would rise through the solution, forming smoke rings, and ignite with little explosions as they reached the surface.

For burning, one used a very long-stemmed "deflagrating" spoon, which one could lower gingerly, with its thimbleful of combustible, into a cylinder of air, or oxygen, or chlorine, or whatever. The flames were all better and brighter if one used oxygen. If one melted sulfur and then lowered it into the oxygen, it took fire and burned with a bright blue flame, producing pungent, titillating, but suffocating sulfur dioxide. Steel wool, purloined from the kitchen, was surprisingly inflammable—this, too, burned brilliantly in oxygen, producing showers of sparks like the sparklers on Guy Fawkes night, and a dirty brown dust of iron oxide.

With chemistry such as this, one was playing with fire, in the literal as well as the metaphorical sense. Huge energies, plutonic forces, were being unleashed, and I had a thrilling but precarious sense of being in control—sometimes just. This was especially so with the intensely exothermic reactions of aluminum and magnesium; they could be used to reduce metallic ores, or even to produce elemental silicon from sand, but a little carelessness, a miscalculation, and one had a bomb on one's hands.

Chemical exploration, chemical discovery, was all the more romantic for its dangers. I felt a certain boyish glee in playing with these dangerous substances, and I was struck, in my reading, by the range of accidents that had befallen the pioneers. Few naturalists had been devoured by wild animals or stung to death by noxious plants or insects; few physicists had lost their eyesight gazing at the heavens, or broken a leg on an inclined plane; but many chemists had lost their eyes, limbs, and even their lives, usually through producing inadvertent explosions or toxins. All the early investigators of phosphorus had burned themselves severely. Bunsen, investigating cacodyl cyanide, lost his right eye in an explosion, and very nearly his life. Several later experimenters, like Moissan, trying to make diamond from graphite in intensely heated, high-pressure "bombs," threatened to blow themselves and their fellow workers to kingdom come.

Humphry Davy, one of my particular heroes, had been nearly asphyxiated by nitrous oxide, poisoned himself with nitrogen peroxide, and severely inflamed his lungs with hydrofluoric acid. Davy also experimented with the first "high" explosive, nitrogen trichloride, which had cost many people fingers and eyes. He discovered several new ways of making the combination of nitrogen and chlorine, and caused a violent explosion on one occasion while he was visiting a friend. Davy himself was partially blinded, and did not recover fully for another four months. (We were not told what damage was done to his friend's house.)

The Discovery of the Elements devoted an entire section to "The Fluorine Martyrs." Although elemental chlorine had been isolated from hydrochloric acid in the 1770s, its far more active cousin, fluorine, was not so easily obtained. All the early experimenters, I read, "suffered the frightful torture of hydrofluoric acid poisoning," and at least two of them died in the process. Fluorine was only isolated in 1886, after almost a century of dangerous trying.

I was fascinated by reading this history, and immediately, recklessly, wanted to obtain fluorine for myself. Hydrofluoric acid was easy to get: Uncle Tungsten used vast quantities of it to "pearl" his lightbulbs, and I had seen great carboys of it in his factory in Hoxton. But when I told my parents the story of the fluorine martyrs, they forbade me to experiment with it in the house. (I compromised by keeping a small gutta-percha bottle of hydrofluoric acid in my lab, but my own fear of it was such that I never actually opened the bottle.)

It was really only later, when I thought about it, that I became astonished at the nonchalant way in which Griffin (and my other books) proposed the use of intensely poisonous substances. I had not the least difficulty getting potassium cyanide from the chemist's, the pharmacy, down the road—it was normally used for collecting insects in a killing bottle—but I could rather easily have killed myself with the stuff. I gathered, over a couple of years, a variety of chemicals that could have poisoned or blown up the entire street, but I was careful—or lucky.[3]

If my nose was stimulated in the lab by certain smells—the pungent, irritating smell of ammonia or sulfur dioxide, the odious smell of hydrogen sulfide—it was much more pleasantly stimulated by the garden outdoors and the kitchen, with its food smells, and its essences and spices, inside. What gave coffee its aroma? What were the essential substances in cloves, apples, roses? What gave onions and garlic and radishes their pungent smell? What, for that matter, gave rubber its peculiar odor? I especially liked the smell of hot rubber, which seemed to me to have a slightly human smell (both rubber and people, I learned later, contain odoriferous isoprene). Why did butter and milk acquire sour smells if they "went off," as they tended to do in hot weather? What gave "turps," oil of turpentine, its lovely, piney smell? Besides all these "natural" smells, there were the smells of

---

[3] Now, of course, none of these chemicals can be bought, and even school or museum laboratories are increasingly confined to reagents that are less hazardous—and less fun.

Linus Pauling, in an autobiographical sketch, described how he, too, obtained potassium cyanide (for a killing bottle) from a local druggist:

> Just think of the differences today. A young person gets interested in chemistry and is given a chemical set. But it doesn't contain potassium cyanide. It doesn't even contain copper sulfate or anything else interesting because all the interesting chemicals are considered dangerous substances. Therefore, these budding young chemists don't have a chance to do anything engrossing with their chemistry sets. As I look back, I think it is pretty remarkable that Mr. Ziegler, this friend of the family, would have so easily turned over one-third of an ounce of potassium cyanide to me, an eleven-year-old boy.

When I paid a visit not long ago to the old building in Finchley which had been Griffin & Tatlock's home a half century ago, it was no longer there. Such shops, such suppliers, which had provided chemicals and simple apparatus and unimaginable delights for generations, have now all but vanished.

the alcohol and acetone that my father used in the surgery, and of the chloroform and ether in my mother's obstetric bag. There was the gentle, pleasant, medical smell of iodoform, used to disinfect cuts, and the harsh smell of carbolic acid, used to disinfect lavatories (it carried a skull and crossbones on its label).

Scents could be distilled, it seemed, from all parts of a plant — leaves, petals, roots, bark. I tried to extract some fragrances by steam distillation, gathering rose petals and magnolia blossoms and grass cuttings from the garden and boiling them with water. Their essential oils would be volatilized in the steam and settle on top of the distillate as it cooled (the heavy, brownish essential oil of onions or garlic, though, would sink to the bottom). Alternatively one could use fat — butter fat, chicken fat — to make a fatty extract, a pomatum; or use solvents like acetone or ether. On the whole my extractions were not too successful, but I succeeded in making some reasonable lavender water, and extracting clove oil and cinnamon oil with acetone. The most productive extractions came from my visits to Hampstead Heath, when I gathered large bags of pine needles and made a fine, bracing green oil full of terpenes — the smell reminded me a little of the Friar's Balsam that I would be set to inhale, in steam, whenever I had a cold.

I loved the smell of fruits and vegetables and would savor everything, sniff at it, before I ate. We had a pear tree in the garden, and my mother would make a thick pear nectar from its fruit, in which the smell of pears seemed heightened. But the scent of pears, I had read, could be made artificially, too (as was done with "pear drops"), without using any pears. One had only to start with one of the alcohols — ethyl, methyl, amyl, whatever — and distill it with acetic acid to form the corresponding ester. I was amazed that something as simple as ethyl acetate could be responsible for the complex, delicious smell of pears, and that tiny chemical changes could transform this to other fruity scents — change the ethyl to isoamyl, and one had the smell of ripening apples; other small modifications would give esters that smelled of bananas or apricots or pineapples or grapes. This was my first experience of the power of chemical synthesis.

There were, besides the pleasant fruity smells, a number of vile, animally smells that one could easily make from simple ingredients or extract from plants. Auntie Len, with her botanical knowledge, sometimes colluded with me here, and introduced me to a plant called stinking goosefoot, a species of *Chenopodium*. If this was distilled in an alkaline medium—I used soda—a particularly vile-smelling and volatile material came off, which stank of rotten crabs or fish. The volatile substance, trimethylamine, was surprisingly simple—I had thought the smell of rotting fish would have a more complex basis. In America, Len told me, they had a plant called skunk cabbage, and this contained compounds that smelled like corpses or putrefying flesh; I asked if she could get me some, but, perhaps fortunately, she could not.

Some of these stinks incited me to pranks. We would get fresh fish every Friday, carp and pike, which my mother would grind to make the gefilte fish for shabbas. One Friday I added a little trimethylamine to the fish, and when my mother smelled this, she grimaced and threw the lot away.

My interest in smells made me wonder how we recognized and categorized odors, how the nose could instantly delineate esters from aldehydes, or recognize a category such as terpenes, as it were, at a glance. Poor as our sense of smell was compared to a dog's—our dog, Greta, could detect her favorite foods if a tin was opened at the other end of the house—there nevertheless seemed in humans to be a chemical analyzer at work at least as sophisticated as the eye or the ear. There did not seem to be any simple order, like the scale of musical tones, or the colors of the spectrum; yet the nose was quite remarkable in making categorizations that corresponded, in some way, to the basic structure of the chemical molecules. All the halogens, while different, had halogenlike smells. Chloroform smelled exactly like bromoform and (while not identical) had the same sort of smell as carbon tetrachloride (sold as the dry-cleaning fluid Thawpit). Most esters were fruity; alcohols—the simplest ones, anyway—had similar "alcoholic" smells; and aldehydes and ketones, too, had their own characteristic smells.

(Errors, surprises, could certainly occur, and Uncle Dave told me how phosgene, carbonyl chloride, the terrible poison gas used in the First World War, instead of signaling its danger by a halogenlike smell, had a deceptive scent like new-mown hay. This sweet, rustic smell, redolent of the hayfields of their boyhood, was the last sensation phosgene-gassed soldiers had just before they died.)

The bad smells, the stenches, always seemed to come from compounds containing sulfur (the smells of garlic and onion were simple organic sulfides, as closely related chemically as they were botanically), and these reached their climax in the sulfuretted alcohols, the mercaptans. The smell of skunks was due to butyl mercaptan, I read—this was pleasant, refreshing, when very dilute, but appalling, overwhelming, at close quarters. (I was delighted, when I read *Antic Hay* a few years later, to find that Aldous Huxley had named one of his less delectable characters Mercaptan.)

Thinking of all the malodorous sulfur compounds and the atrocious smell of selenium and tellurium compounds, I decided that these three elements formed an olfactory as well as a chemical category, and thought of them thereafter as the "stinkogens."

I had smelled a bit of hydrogen sulfide in Uncle Dave's lab— it smelled of rotten eggs and farts and (I was told) volcanoes. A simple way of making it was to pour dilute hydrochloric acid on ferrous sulfide. (The ferrous sulfide, a great chunky mass of it, I made myself by heating iron and sulfur together till they glowed and combined.) The ferrous sulfide bubbled when I poured hydrochloric acid on it, and instantly emitted a huge quantity of stinking, choking hydrogen sulfide. I threw open the doors into the garden and staggered out, feeling very queer and ill, remembering how poisonous the gas was. Meanwhile, the infernal sulfide (I had made a lot of it) was still giving off clouds of toxic gas, and this soon permeated the house. My parents were, by and large, amazingly tolerant of my experiments, but they insisted, at this point, on having a fume cupboard installed and on my using, for such experiments, less generous quantities of reagents.

When the air had cleared, morally and physically, and the fume cupboard had been installed, I decided to make other gases, simple compounds of hydrogen with other elements besides sulfur. Knowing that selenium and tellurium were closely akin to sulfur, in the same chemical group, I employed the same basic formula: compounding the selenium or tellurium with iron, and then treating the ferrous selenide or ferrous telluride with acid. If the smell of hydrogen sulfide was bad, that of hydrogen selenide was a hundred times worse—an indescribably horrible, disgusting smell that caused me to choke and tear, and made me think of putrefying radishes or cabbage (I had a fierce hatred of cabbage and brussels sprouts at this time, for boiled, overboiled, they had been staples at Braefield).

Hydrogen selenide, I decided, was perhaps the worst smell in the world. But hydrogen telluride came close, was also a smell from hell. An up-to-date hell, I decided, would have not just rivers of fiery brimstone, but lakes of boiling selenium and tellurium, too.

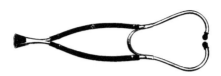

9

HOUSECALLS

My father was not given to emotion or intimacy, at least in the context, the confines, of the family. But there were certain times, precious times, when I did feel close to him. I have very early memories of seeing him reading in our library, and his concentration was such that nothing could disturb him, for everything outside the circle of his lamp was completely tuned out of his mind. For the most part he read the Bible or the Talmud, though he also had a large collection of books on Hebrew, which he spoke fluently, and Judaism—the library of a grammarian and scholar. Seeing his intense absorption in reading, and the expressions that would appear on his face as he read (an involuntary smile, a grimace, a look of perplexity or delight), perhaps drew me to reading very early myself, so that even before the war I would sometimes join him in the library, reading my book alongside him, in a deep but unspoken companionship.

If there were no housecalls to do in the evening, my father would settle down after dinner with a torpedo-shaped cigar. He would palpate it gently, then hold it to his nose to test its aroma and freshness, and if it was satisfactory he would make a V-shaped incision in its tip with his cutter. He would light it carefully with a long match, rotating it so that it lit evenly. The tip would glow red as he drew, and his first exhalation was a sigh of satisfaction. He would puff away gently as he read, and the air would turn blue and opalescent with smoke, enfolding us both in a fragrant cloud. I loved the smell of the beautiful Havanas he smoked, and loved to watch the grey cylinder of ash grow longer and longer, wondering how long it would get before it dropped on his book.

I felt closest to him, truly his son, when we went swimming together. My father's passion, from an early age, had been swimming (as his father had been a swimmer before him), and he had been a swimming champion when he was younger, having won the fifteen-mile race off the Isle of Wight three years in succession. He had introduced each of us to the water when we were babies, taking us to the Highgate Ponds in Hampstead Heath.

The slow, measured, mile-eating stroke he had was not entirely suited to a little boy. But I could see how my old man, huge and cumbersome on land, became transformed—graceful, like a porpoise—in the water; and I, self-conscious, nervous, and also rather clumsy, found the same delicious transformation in myself, found a new being, a new mode of being, in the water. I have a vivid memory of a summer holiday at the seaside, the month after my fifth birthday, when I ran into my parents' room and tugged at the great whalelike bulk of my father. "Come on, Pop!" I said. "Let's go for a swim." He turned over slowly and opened one eye: "What do you mean, waking an old man of forty-three like this at six in the morning?" Now that my father is dead, and I myself am in my sixties, this memory of so long ago tugs, makes me equally want to laugh and cry.

Later we would swim together in the large open-air pool in Hendon, or the Welsh Harp on Edgware Road, a small lake (I was never sure whether it was natural or artificial) where my father had once kept a boat. After the war, as a twelve-year-old, I could begin to match his strokes, and maintain the same rhythm, swimming in unison with him.

I sometimes went along with my father on housecalls on Sunday mornings. He loved doing housecalls more than anything else, for they were social and sociable as well as medical, would allow him to enter a family and home, get to know everybody and their circumstances, see the whole complexion and context of a condition. Medicine, for him, was never just diagnosing a disease, but had to be seen and understood in the context of patients' lives, the particularities of their personalities, their feelings, their reactions.

He would have a typed list of a dozen patients and their addresses, and I would sit next to him in the front seat of the car while he told me, in very human terms, what each patient had. When he arrived, I would get out with him, allowed, usually, to carry his medical bag. Sometimes I would go into the sickroom with him and sit quietly while he questioned and examined a patient—a questioning and examining which seemed swift and light, and yet one that reached depths and exposed for him the origins of each illness. I loved to see him percuss the chest, tapping it delicately but powerfully with his strong stubby fingers, feeling, sensing, the organs and their state beneath. Later, when I became a medical student myself, I realized what a master of percussion he was, and how he could tell more by palpating and percussing and listening to a chest than most doctors could from an X-ray.

At other times, if the patient was very ill, or contagious, I would sit with the family in their kitchen or dining room. After my father had seen the patient upstairs, he would come down, wash his hands carefully, and make for the kitchen. He loved to eat, and he knew the contents of the refrigerators in all his patients' houses—and the families seemed to enjoy giving the good doctor food. Seeing patients, meeting families, enjoying himself, eating, were all inseparable in the medicine he practiced.

Driving through the City, deserted on a Sunday, was a sobering experience in 1946, for the devastation wrought by the bombing was still fresh, and there had been little rebuilding as yet. This was even more evident in the East End, where a fifth of the buildings, perhaps, had been leveled. But there was still a strong Jewish community there, and restaurants and delicatessens like no others in the world. My father had qualified at the London Hospital in Whitechapel Road, and as a young man had been the Yiddish-speaking doctor of the Yiddish-speaking community around it for ten years. He looked back on these early days with peculiar affection. We would sometimes visit his old surgery in New Road — it was here that all my brothers had been born, and a physician nephew, Neville, now practiced.

We would walk up and down "the Lane," that section of Petticoat Lane between Middlesex Street and Commercial Street, where all the stallholders hawked their wares. My parents had left the East End in 1930, but my father still knew many of the hawkers by name. Jabbering with them, reverting to the Yiddish of his youth, my old father (what do I mean "old"? I am now fifteen years older than the fifty-year-old he was then) became boyish, rejuvenated, showed an earlier, more alive self that I normally did not see.

We would always go to Marks of the Lane, where one could buy a latke for sixpence, and the best smoked salmon and herrings in London, salmon of an unbelievable melting softness which made it one of the few, genuinely paradisal experiences on this earth.

My father had always had a very robust appetite, and the strudel and herrings at his patients' houses, and the latkes at Marks, were, in his mind, just preludes to the real meal. There were a dozen superb kosher restaurants within a few blocks, each with its own incomparable specialties. Should it be Bloom's on Aldgate, or Ostwind's, where one could enjoy the marvelous smells of the basement bakery wafting upstairs? Or Strongwater's, where there was a particular sort of kreplach, *varenikas*, to which my father was dangerously addicted? Usually, however, we would end up at Silberstein's, where, in addition to the meat restaurant downstairs, there was a dairy restaurant, with wonderful milky soups and fish, upstairs. My father adored carp, in particular, and would suck at the fish heads, noisily, with great gusto.

Pop was a calm, unflappable driver when he went on his housecalls—he had a sedate, rather slow Wolseley at the time, appropriate to the petrol rationing still in force—but before the war there was a very different side to him. His car then was an American one, a Chrysler, with a raw power and a turn of speed unusual in the 1930s. He also had a motorcycle, a Scott Flying Squirrel, with a two-stroke, 600 cc, watercooled engine, and a high-pitched exhaust like a scream. It developed nearly thirty horsepower, and was much more akin, he liked to say, to a flying horse. He loved to take off on this if he had a free Sunday morning, eager to shake off the city and give himself to the wind and the road, his practice, his cares forgotten for a while. Sometimes I had dreams in which I was riding or flying the bike myself, and I determined to get one when I was grown up.

When T. E. Lawrence's *The Mint* came out in 1955, I read my father a piece, "The Road," Lawrence had written about his motorbike (by this time I had a bike, a Norton, myself):

> A skittish motor-bike with a touch of blood in it is better than all the riding animals on earth, because of its logical extension of our faculties, and the hint, the provocation, to excess . . .

My father smiled and nodded in agreement, as he thought back to his own biking days.

My father had originally wondered about an academic career in neurology, and had been a houseman, an intern (along with Jonathan Miller's father), to Sir Henry Head, the famous neurol-

95

ogist, at the London Hospital. At this point, Head himself, still at the height of his powers, had developed Parkinson's disease, and this, my father said, would sometimes cause him to run involuntarily, or festinate, the length of the old neurology ward, so that he would have to be caught by one of his own patients. When I had difficulty imagining what this was like, my father, an excellent mimic, imitated Head's festination, careering down Exeter Road at an ever-accelerating pace, and getting me to catch him. Head's own predicament, my father thought, made him especially sensitive to the predicaments of his patients, and I think my father's imitations—he could imitate asthma, convulsions, paralyses, anything—springing from his vivid imagination of what it was like for others, served the same purpose.

When it was time for my father to open his own practice, he decided, despite this early training in neurology, that general practice would be more real, more "alive." Perhaps he got more than he bargained for, for when he opened his practice in the East End in September 1918, the great influenza epidemic was just getting started. He had seen wounded soldiers when he was a houseman at the London, but this was nothing to the horror of seeing people in paroxysms of coughing and gasping, suffocating from the fluid in their lungs, turning blue and dropping dead in the streets. A strong, healthy young man or woman, it was said, could die from the flu within three hours of getting it. In those three desperate months at the end of 1918, the flu killed more people than the Great War itself had, and my father, like every doctor at the time, found himself overwhelmed, sometimes working forty-eight hours at a stretch.

At this point he engaged his sister Alida—a young widow with two children who had returned to London from South Africa three years before—to work as his assistant in the dispensary. Around the same time, he took on another young doctor, Yitzchak Eban, to help him on his rounds. Yitzchak had been born in Joniški, the same little village in Lithuania where the

Sacks family lived. Alida and Yitzchak had been playmates as infants, but then in 1895 his family had gone to Scotland, a few years before the Sackses had come to London. Reunited twenty years later, working together in the febrile and intense atmosphere of the epidemic, Alida and Yitzchak fell in love, and married in 1920.

As children, we had relatively little contact with Auntie Alida (though I thought of her as the quickest and wittiest of my aunts—she had sudden intuitions, sudden swoops of thought and feeling, which I came to think of as characteristic of the "Sacks mind," in contrast to the more methodical, more analytical, mental processes of the Landaus). But Auntie Lina, my father's eldest sister, was a constant presence. She was fifteen years older than Pop, tiny in size—four foot nine in her high heels—but with an iron will, a ruthless determination. She had dyed golden hair, as coarse as a doll's, and gave off a mixed scent of garlic, sweat, and patchouli. It was Lina who had furnished our house, and Lina who would often provide us at 37 with certain special items which she herself cooked—fish cakes (Marcus and David called her Fishcake, or sometimes Fishface, after these), rich crumbly cheesecakes, and, at Passover, matzoh balls of an incredible tellurian density, which would sink like little planetismals below the surface of the soup. Careless of the social graces, she would bend down at the table, when at home, and blow her nose on the tablecloth. Despite this, she was enchanting in company, when she would glitter and coquette, but also listen intently, judging the character and motive of everyone around her. She would draw confidences out of the unwary, and with her diabolical memory, retain all that she had heard.[1]

But her ruthlessness, her unscrupulousness, had a noble purpose, for she used them to raise money for the Hebrew University

---

[1] Many years later, when I read Keynes's wonderful description of Lloyd George (in *The Economic Consequences of the Peace*), I was strangely reminded of

in Jerusalem. She had dossiers, it seemed, on everyone in England, or so I sometimes imagined, and once she was certain of her information and sources, she would lift the phone. "Lord G.? This is Lina Halper." There would be a pause, a gasp, Lord G. would know what was coming. "Yes," she would continue pleasantly, "yes, you know me. There is that little business—no, we won't go into details—that little affair in Bognor, in March '23. . . . No, of course I won't mention it, it'll be our little secret—what can I put you down for? Fifty thousand, perhaps? I can't tell you what it would mean to the Hebrew University." By this sort of blackmail Lina raised millions of pounds for the university, the most efficient fund-raiser, probably, they had ever known.

Lina, considerably the oldest, had been "a little mother" to her much younger siblings when they came to England from Lithuania in 1899, and after the early death of her husband, she took over my father, in a sense, and vied with my mother for his company and affections. I was always aware of the tension, the unspoken rivalry, between them, and had a sense of my father—soft, passive, indecisive—being pulled this way and that between them.

While Lina was regarded by many in the family as a sort of monster, she had a soft spot for me, as I had for her. She was especially important to me, to all of us perhaps, at the start of the war, for we were in Bournemouth on our summer holiday when war

---

Auntie Lina. Keynes speaks of the British prime minister's "unerring, almost medium-like sensibility to everyone immediately around him."

To see [him], watching the company with six or seven senses not available to ordinary men, judging character, motive, and sub-conscious impulse, perceiving what each was thinking, and even what each was going to say next, compounding with telepathic instinct the argument or appeal best suited to the vanity, weakness, or self-interest of his immediate auditor was to realize that the poor President [Wilson] would be playing blind man's buff in that party.

was declared, and our parents, as doctors, had to leave immediately for London, leaving the four of us with the nanny. They came back a couple of weeks later, and my relief, our relief, was prodigious. I remember rushing down the garden path when I heard the hoot of the car, and flinging myself bodily into my mother's arms, so vehemently I almost knocked her over. "I've missed you," I cried, "I've missed you so much." She hugged me, a long hug, holding me tight in her arms, and the sense of loss, of fear, suddenly dissolved.

Our parents promised to come again very soon. They would try to manage the next weekend, they said, but there was a great deal for them to do in London—my mother was occupied with emergency trauma surgery, my father was organizing local G.P.s for casualties in air raids. But this time they did not come at the weekend. Another week passed, and another, and another, and something, I think, broke inside me at this point, for when they did come again, six weeks after their first visit, I did not run up to my mother or embrace her as I had the first time, but treated her coldly, impersonally, like a stranger. She was, I think, shocked and bewildered by this, but did not know how to bridge the gulf that had come between us.

At this point, when the effects of parental absence had become unmistakable, Lina came up, took over the house, did the cooking, organized our lives, and became a little mother to us all, filling in the gap left by our own mother's absence.

This little interlude did not last long—Marcus and David went off to medical school, and Michael and I were packed off to Braefield. But I never forgot Lina's tenderness to me at this time, and after the war I took to visiting her in London, in her high-ceilinged, brocaded room in Elgin Avenue. She would give me cheesecake, sometimes a fish cake, and a little glass of sweet wine, and I would listen to her reminiscences of the old country. My father was only three or four when he left, and had no memories of it; Lina, eighteen or nineteen at the time, had vivid and fascinating memories of Joniški, the shtetl near Vilna where they had all been born, and of her parents, my grandparents, as they were in comparative youth. It may be that she had a special feeling for me as the youngest, or because I had the same name as her father, Elivelva, Oliver Wolf. I had the sense, too, that she was lonely and enjoyed the visits of her young nephew.

Then there was my father's brother, Bennie. Bennie had been excommunicated, left the family fold, at nineteen, when he had gone to Portugal and married a gentile, a shiksa. This was a crime so scandalous, so heinous in the eyes of the family that his name was never mentioned thereafter. But I knew there was something hidden, a family secret of sorts; I surprised certain silences, certain awkwardnesses, sometimes, when my parents whispered together, and I once saw a photo of Bennie on one of Lina's embossed cabinets (she said it was someone else, but I picked up the hesitation in her voice).

My father, always powerfully built, started to put on weight after the war and decided to go at regular intervals to a fat farm in Wales. These visits never seemed to do him much good, weightwise, but he would come back from them looking happy and well, his London pallor replaced by a healthy tan. It was only after his death, many years later, that, looking through his papers, I found a sheaf of plane tickets that told the true story— he had never been to the fat farm at all, but loyally, secretly, had been going to visit Bennie in Portugal all these years.

10

A CHEMICAL LANGUAGE

Uncle Dave saw all science as a wholly human, no less than an intellectual and technological, enterprise, and it seemed natural to me, in my turn, to do the same. When I set up my lab and started some chemical experiments of my own, I wanted to learn about the history of chemistry in a more general way, to find out what chemists did, how they thought, the atmosphere in centuries past. I had long been fascinated by our family and family tree—by tales of the uncles who had gone off to South Africa, and of the man who had fathered them all, and of the first ancestor of my mother's of whom we had any record, an alchemically inclined rabbi, it was said, one Lazar Weiskopf, who lived in Lübeck in the seventeenth century. This may have been the incitement to a more general love of history, and a tendency, perhaps, to see it in familial terms. And so the scientists, the early chemists, whom I read about became, in a sense, honorary ancestors, people to whom, in fantasy, I had a sort of connection. I needed to understand how these early chemists thought, to imagine myself into their worlds.

Chemistry as a true science, I read, made its first emergence with the work of Robert Boyle in the middle of the seventeenth century. Twenty years Newton's senior, Boyle was born at a time when the practice of alchemy still held sway, and he still maintained a variety of alchemical beliefs and practices, side by side with his scientific ones. He believed that gold could be created, and that he had succeeded in creating it (Newton, also an alchemist, advised him to keep silent about this). He was a man of immense curiosity (of "holy curiosity," in Einstein's phrase), for all the wonders of nature, Boyle felt, proclaimed the glory of God, and this led him to examine a huge range of phenomena.

He examined crystals and their structure, and was the first to discover their cleavage planes. He explored color, and wrote a book on this which influenced Newton. He devised the first chemical indicator, a paper soaked with syrup of violets which would turn red in the presence of acid fluids, green with alkaline ones. He wrote the first book in English on electricity. He prepared hydrogen, without realizing it, by putting iron nails in sulfuric acid. He found that although most fluids contracted when frozen, water expanded. He showed that a gas (later realized to be carbon dioxide) was evolved when he poured vinegar on powdered coral, and that flies would die if kept in this "artificial air." He investigated the properties of blood and was interested in the possibility of blood transfusion. He experimented with the perception of odors and tastes. He was the first to describe semipermeable membranes. He provided the first case history of acquired achromatopsia, a total loss of color vision following a brain infection.

All these investigations and many others he described in language of great plainness and clarity, utterly different from the arcane and enigmatic language of the alchemists. Anyone could read him and repeat his experiments; he stood for the openness of science, as opposed to the closed, hermetic secrecy of alchemy.

Although his interests were universal, chemistry seemed to hold a very special appeal for him (even as a youth he called his own chemical laboratory "a kind of Elysium"). He wished, above all, to understand the nature of matter, and his most famous book, *The Sceptical Chymist,* was written to debunk the mystical doctrine of the Four Elements, and to unite the enormous, centuries-old empirical knowledge of alchemy and pharmacy with the new, enlightened rationality of his age.

The ancients had thought in terms of four basic principles or elements—Earth, Air, Fire, and Water. I think these were pretty much my own categories as a five-year-old child (though metals may have made a special, fifth category for me), but I found it less easy to imagine the Three Principles of the alchemists, where "Sulfur" and "Mercury" and "Salt" meant not ordinary sulfur and mercury and salt but "philosophical" Sulfur, Mercury, and Salt: Mercury conferring luster and hardness to a substance, Sulfur conferring color and combustibility, Salt conferring solidity and resistance to fire.

Boyle hoped to replace these ancient, mystical notions of Elements and Principles with a rational and empirical one, and provided the first modern definition of an element:

> I now mean by Elements [he wrote] . . . certain Primitive and Simple, or perfectly unmingled bodies; which not being made up of any other bodies, or of one another, are the ingredients of which all those call'd perfectly mixd Bodies are immediately compounded, and into which they are ultimately resolved.

But since he gave no examples of such "Elements" or of how their "unmingledness" was to be demonstrated, his definition seemed too abstract to be useful.

Though I found *The Sceptical Chymist* unreadable, I was delighted by Boyle's 1660 *New Experiments,* where he set out, with an enchanting vividness and a wealth of personal detail, more than forty experiments using his "Pneumatical Engine" (an air pump that his assistant Robert Hooke had invented), with which he could evacuate much of the air from a closed vessel.[1] In these experiments Boyle effectively demolished the ancient belief that air was an ethereal, all-pervading medium by showing that it was a material substance with physical and chemical properties of its own, that it could be compressed or rarefied or even weighed.

Evacuating the air from a closed vessel that contained a lit candle or a glowing coal, Boyle found that these ceased to burn as the air was rarefied, although the coal would begin to glow again if air was reintroduced—thus showing that air was necessary for combustion. He showed, too, that various creatures—insects, birds, or mice—would become distressed or die if the air pressure was reduced, but might revive when air was readmitted to

---

[1] Hooke himself was to become a marvel of scientific energy and ingenuity, abetted by his mechanical genius and mathematical ability. He kept voluminous, minutely detailed journals and diaries, which provide an incomparable picture not only of his own ceaseless mental activity, but of the whole intellectual atmosphere of seventeenth-century science. In his *Micrographia,* Hooke illustrated his compound microscope, along with drawings of the intricate, never-before-seen structures of insects and other creatures (including a famous picture of a Brobdingnagian louse, attached to a human hair as thick as a barge pole). He judged the frequency of flies' wingbeats by their musical pitch. He interpreted fossils, for the first time, as the relics and impressions of extinct animals. He illustrated his designs for a wind gauge, a thermometer, a hygrometer, a barometer. And he showed an intellectual audacity sometimes even greater than Boyle's, as with his understanding of combustion, which, he said, "is made by a substance inherent, and mixt with the Air." He identified this with "that property in the Air which it loses in the Lungs." This notion of a substance present in limited amounts in the air that is required for and gets used up in combustion and respiration is far closer to the concept of a chemically active gas than Boyle's theory of igneous particles.

the vessel. He was struck by this similarity between combustion and respiration.

He investigated whether a bell could be heard through a vacuum (it could not), whether a magnet could exert power through a vacuum (it could), whether insects could fly in a vacuum (this he could not judge, because the insects "swooned" with reduction of air pressure), and he examined the effects of reduced air pressure on the glowing of glowworms (they glowed less brightly).

I loved reading about these experiments and tried repeating some of them for myself—our Hoover was a good substitute for Boyle's air pump. I loved the playfulness of the whole book, so different from the philosophical dialogues in *The Sceptical Chymist*. (Indeed, Boyle himself was not unaware of this: "I disdain not to take notice even of ludicrous experiments, and think that the plays of boys may sometimes deserve to be the study of philosophers.")

Boyle's personality appealed to me greatly, as did his omnivorous curiosity, his fondness for anecdote, and his occasional puns (as when he wrote that he preferred to work on things "luciferous rather than lucriferous"). I could imagine him as a person, and a person I would like, despite the gulf of three centuries between us.

Many of Hooke's ideas were almost completely ignored and forgotten, so that one scholar observed in 1803, "I do not know a more unaccountable thing in the history of science than the total oblivion of this theory of Dr. Hooke, so clearly expressed, and so likely to catch attention." One reason for this oblivion was the implacable enmity of Newton, who developed such a hatred of Hooke that he would not consent to assume the presidency of the Royal Society while Hooke was still alive, and did all he could to extinguish Hooke's reputation. But deeper than this is perhaps what Gunther Stent calls "prematurity" in science, that many of Hooke's ideas (and especially those on combustion) were so radical as to be unassimilable, even unintelligible, in the accepted thinking of his time.

. . .

Antoine Lavoisier, born almost a century after Boyle, would become known as the real founder, the father, of modern chemistry. There was already a huge amount of chemical knowledge, chemical sophistication, before his time, some of it bequeathed by the alchemists (for it was they who pioneered the apparatus and techniques of distillation and crystallization and a range of chemical procedures), some of it by apothecaries, and much of it, of course, by early metallurgists and miners.

Yet although a multitude of chemical reactions had been explored, there was no systematic weighing or measurement of these reactions. The composition of water was unknown, as was the composition of most other substances. Minerals and salts were classified by their crystalline form, or other physical properties, rather than their constituents. There was no clear notion of elements or compounds.

There was, moreover, no overall theoretical framework in which chemical phenomena could be placed, only the somewhat mystical theory of phlogiston, which was supposed to explain all chemical transformations. Phlogiston was the principle of Fire. Metals were combustible, it was supposed, because they contained some phlogiston, and when they were burned, the phlogiston was released. When their earths were smelted with charcoal, conversely, the charcoal donated *its* phlogiston and reconstituted the metal. Thus a metal was a sort of composite or "compound" of its earth, its calx, and phlogiston. Every chemical process—not only of smelting and calcination, but the actions of acids and alkalis, and the formation of salts—could be attributed to the addition or removal of phlogiston.

It was true that phlogiston had no visible properties, could not be bottled, demonstrated, or weighed—but after all, was this not equally true of electricity (another great source of mystery and fascination in the eighteenth century)? Phlogiston had an instinctive, poetic, mythic appeal, making fire at once a material and a spirit. But for all its metaphysical roots, the phlogiston theory was the first specifically chemical theory (as opposed to the mechanical, corpuscular one that Boyle had envisaged in the 1660s); it attempted to account for chemical properties and reactions in terms of the presence or absence, or transference, of a specific chemical principle.

It was into this half-metaphysical, half-poetic atmosphere that Lavoisier—hardheaded, keenly analytical and logical, a child of the Enlightenment and an admirer of the Encyclopedists—came of age in the 1770s. By the age of twenty-five, Lavoisier had already done pioneering geological work, shown great chemical and polemical skill (he had written a prizewinning essay on the best means of illuminating a city at night, as well as a study of the setting and binding of plaster of Paris), and been elected to the Academy.[2] But it was in relation to the theory of phlogiston

---

[2] In his biography of Lavoisier, Douglas McKie includes an exhaustive list of Lavoisier's scientific activities which paints a vivid picture of his times, no less than his own remarkable range of mind: "Lavoisier took part," McKie writes,

in the preparation of reports on the water supply of Paris, prisons, mesmerism, the adulteration of cider, the site of the public abattoirs, the newly-invented "aerostatic machines of Montgolfier" (balloons), bleaching, tables of specific gravity, hydrometers, the theory of colors, lamps, meteorites, smokeless grates, tapestry making, the engraving of coats-of-arms, paper, fossils, an invalid chair, a water-driven bellows, tartar, sulfur springs, the cultivation of cabbage and rape seed and the oils extracted thence, a tobacco grater, the working of coal mines, white soap, the decomposition of nitre, the manufacture of starch . . . the storage of fresh water on ships, fixed air, a reported occurrence of oil in spring water . . . the removal of oil and grease from silks and woollens, the preparation of nitrous ether by distillation, ethers, a reverberatory hearth, a new ink and inkpot to which it was only necessary to add water in order to maintain the supply of ink . . . , the estimation of alkali in mineral waters, a powder magazine for the Paris Arsenal, the mineralogy of the Pyrenees, wheat and flour, cesspools and the air arising from them, the alleged occurrence of gold in the ashes of plants, arsenic acid, the parting of gold and silver, the base of Epsom salt, the winding of silk, the solution of tin used in dyeing, volcanoes, putrefaction, fire-that his intellect and ambition became sharply focused. The idea of phlogiston seemed to him metaphysical, insubstantial, and the point of attack, he saw at once, lay in meticulous quantitative experiments with combustion. Did substances indeed decrease in weight when they burned, as one would expect if they lost their phlogiston? Common experience, indeed, suggested that this was so, that substances "burned away"—a candle dwindled in size as it burned, organic substances charred and shriveled, sulfur and charcoal vanished completely, but this did not seem to be the case with regard to the burning of metals.

In 1772 Lavoisier read of the experiments of Guyton de Morveau, who had confirmed in experiments of exceptional precision and care that metals *increased* in weight when they were roasted in air.[3] How could this be reconciled with the notion that something—phlogiston—was lost in burning? Lavoisier found

---

extinguishing liquids, alloys, the rusting of iron, a proposal to use "inflammable air" in a public firework display (this at the request of the police), coal measures, dephlogisticated marine acid, lamp wicks, the natural history of Corsica, the mephitis of the Paris wells, the alleged solution of gold in nitric acid, the hygrometric properties of soda, the iron and salt works of the Pyrenees, argentiferous lead mines, a new kind of barrel, the manufacture of plate glass, fuels, the conversion of peat into charcoal, the construction of corn mills, the manufacture of sugar, the extraordinary effects of a thunder bolt, the retting of flax, the mineral deposits of France, plated cooking vessels, the formation of water, the coinage, barometers, the respiration of insects, the nutrition of vegetables, the proportion of the components in chemical compounds, vegetation, and many other subjects, far too many to be described here, even in the briefest terms.

[3] Boyle had experimented with the burning of metals a hundred years before, and was well aware that these increased in weight when burned, forming a calx or ash that was heavier than the original. But his explanations of the increase of weight were mechanical, not chemical: he saw it as the absorption of "particles of fire." Similarly, he saw air itself not in chemical terms, but rather as an elastic fluid of a peculiar sort, used in a sort of mechanical ventilation, to wash the impurities out of the lungs. Findings were not consistent in the century that followed Boyle, partly because the gigantic "burning glasses" used

Guyton's explanation—that phlogiston had "levity" and buoyed up the metals that contained it—absurd. But Guyton's impeccable results nonetheless incited Lavoisier as nothing had before. It was, like Newton's apple, a fact, a phenomenon, that demanded a new theory of the world.

The work before him, he wrote, "seemed to me destined to bring about a revolution in physics and in chemistry. I have felt bound to look upon all that has been done before me merely as suggestive . . . like separate pieces of a great chain." It remained for someone, for *him*, he felt, to join all the links of the chain with "an immense series of experiments . . . in order to lead to a continuous whole" and to form a theory.

While confiding this grandiose thought to his lab notebook, Lavoisier set to systematic experiments, repeating many of his predecessors' work, but this time using a closed apparatus and meticulously weighing everything before and after the reaction, a procedure which Boyle, and even the most meticulous chemists of Lavoisier's own time, had neglected. Heating lead and tin in closed retorts until they were converted to ash, he was able to show that the total weight of his reactants neither increased nor decreased during a reaction. Only when he broke open his retorts, allowing air to rush in, did the weight of the ash increase—and by exactly the same amount as the metals themselves had increased in being calcined. This increase, Lavoisier felt, must be due to the "fixation" of air, or some part of it.

In the summer of 1774, Joseph Priestley, in England, found that when he heated red calx of mercury (mercuric oxide) it gave off an "air" which, to his amazement, seemed even stronger or purer than common air.

---

were of such power as to cause some metallic oxides to partly vaporize or sublime, causing losses rather than increases in weight. But even more frequently there was no weighing at all, for analytical chemistry, at this point, was still largely qualitative.

A candle burned in this air [he wrote] with an amazing strength of flame; and a bit of red hot wood crackled and burned with a prodigious rapidity, exhibiting an appearance something like that of iron glowing with a white heat, and throwing out sparks in all directions.

Entranced, Priestley had investigated this further, and found that mice could live in this air four or five times longer than in ordinary air. And being thus convinced that his new "air" was benign, he tried it himself:

The feeling of it to my lungs was not sensibly different from that of common air; but I fancied that my breast felt peculiarly light and easy for some time afterwards. Who can tell but that, in time, this pure air may become a fashionable article in luxury. Hitherto only two mice and myself have had the privilege of breathing it.

In October of 1774, Priestley went to Paris and spoke of his new "dephlogisticated air" to Lavoisier. And Lavoisier saw in this what Priestley himself did not: the vital clue to what had perplexed and eluded him, the real nature of what was happening in combustion and calcination.[4] He repeated Priestley's experiments, amplified, quantified, refined them. Combustion, it was now clear to him, was a process involving not the loss of a substance (phlogiston), but the combination of the combustible

---

[4] In this same month, Lavoisier got a letter from Scheele describing the preparation of what Scheele called Fire Air (oxygen) admixed with Fixed Air (carbon dioxide), from heating silver carbonate; Scheele had obtained pure Fire Air from mercuric oxide, even before Priestley had. But in the event, Lavoisier claimed the discovery of oxygen for himself and scarcely acknowledged the discoveries of his predecessors, feeling that they did not realize what it was that they had observed.

All this, and the question of what constitutes "discovery," is explored in the play *Oxygen,* by Roald Hoffmann and Carl Djerassi.

material with a part of atmospheric air, a gas, for which he now coined the term *oxygen*.[5]

Lavoisier's demonstration that combustion was a chemical process—oxidation, as it could now be called—implied much else, and was for him only a fragment of a much wider vision, the revolution in chemistry that he had envisaged. Roasting metals in closed retorts, showing that there was no ghostly weight gain from "particles of fire" or weight loss from loss of phlogiston, had demonstrated to him that there was neither creation nor loss of matter in such processes. This principle of conservation, moreover, applied not only to the total mass of products and reactants, but to each of the individual elements involved. When one fermented sugar with yeast and water in a closed vessel to yield alcohol, as in one of his experiments, the total amounts of carbon and hydrogen and oxygen always stayed the same. They might be reaggregated chemically, but their amounts were unchanged.

The conservation of mass implied a constancy of composition and decomposition. Thus Lavoisier was led to define an element as a material that could not be decomposed by existing means, and this enabled him (with de Morveau and others) to draw up a list of genuine elements—thirty-three distinct, undecomposable, elementary substances, replacing the four Elements of the ancients.[6] This in turn allowed Lavoisier to draw up a "balance

---

[5] Replacing the concept of phlogiston with that of oxidation had immediate practical effects. It was now clear, for example, that a burning fuel needed as much air as possible for complete combustion. François-Pierre Argand, a contemporary of Lavoisier's, was quick to exploit the new theory of combustion, designing a lamp with a flat ribbon wick, bent to fit inside a cylinder, so that air could reach it from both the inside and the outside, and a chimney which produced an updraft. The Argand burner was well established by 1783; there had been no lamp so efficient or so brilliant before.

[6] Lavoisier's list of elements included the three gases he had named (oxygen, azote [nitrogen], and hydrogen), three nonmetals (sulfur, phosphorus, and car-

sheet," as he called it, a precise accounting of each element in a reaction.

The language of chemistry, Lavoisier now felt, had to be transformed to go with his new theory, and he undertook a revolution of nomenclature, too, replacing the old, picturesque but uninformative terms—like butter of antimony, jovial bezoar, blue vitriol, sugar of lead, fuming liquor of Libavius, flowers of zinc—with precise, analytic, self-explanatory ones. If an element was compounded with nitrogen, phosphorus, or sulfur, it became a nitride, a phosphide, a sulfide. If acids were formed, through the addition of oxygen, one might speak of nitric acid, phosphoric acid, sulfuric acid; and of the salts of these as nitrates, phosphates, and sulfates. If smaller amounts of oxygen were present, one might speak of nitrites or phosphites instead of nitrates and phosphates, and so on. Every substance, elementary or compound, would have its true name, denoting its composition and chemical character, and such names, manipulated as in an algebra, would instantly indicate how they might interact or behave in different circumstances. (Although I was keenly conscious of the advantages of the new names, I missed the old ones, too, for they had a poetry, a strong feeling of their sensory qualities or hermetic antecedents, which was entirely missing from the new, systematic and scentless chemical names.)

Lavoisier did not provide symbols for the elements, nor did he use chemical equations, but he provided the essential background to these, and I was thrilled by his notion of a balance sheet, this algebra of reality, for chemical reactions. It was like

---

bon), and seventeen metals. It also included muriatic, fluoric, and boracic "radicals" and five "earths": chalk, magnesia, baryta, alumina, and silica. These radicals and earths, he divined, were compounds containing new elements, which he thought would soon be obtained (all of them were indeed obtained by 1825, except fluorine, which defeated isolation for another sixty years). His final two "elements" were Light and Heat—as if he had not been wholly able to free himself from the specter of phlogiston.

seeing language, or music, written down for the first time. Given this algebraic language, one might not need an actual afternoon in the lab—one could in effect do chemistry on a blackboard, or in one's head.

All of Lavoisier's enterprises—the algebraic language, the nomenclature, the conservation of mass, the definition of an element, the formation of a true theory of combustion—were organically interlinked, formed a single marvelous structure, a revolutionary refounding of chemistry such as he had dreamed of, so ambitiously, in 1773. The path to his revolution was not easy or direct, even though he presents it as obvious in the *Elements of Chemistry;* it required fifteen years of genius time, fighting his way through labyrinths of presupposition, fighting his own blindnesses as he fought everyone else's.

There had been violent disputes and conflicts during the years in which Lavoisier was slowly gathering his ammunition, but when the *Elements* was finally published—in 1789, just three months before the French Revolution—it took the scientific world by storm. It was an architecture of thought of an entirely new sort, comparable only to Newton's *Principia.* There were a few holdouts—Cavendish and Priestley were the most eminent of these—but by 1791 Lavoisier could say, "all young chemists adopt the theory and from that I conclude that the revolution in chemistry has come to pass."

Three years later Lavoisier's life was ended, at the height of his powers, on the guillotine. The great mathematician Lagrange, lamenting the death of his colleague and friend, said: "It required only a moment to sever his head, and one hundred years, perhaps, may not suffice to produce another like it."

Reading of Lavoisier and the "pneumatic" chemists who preceded him stimulated me to experiment more with heating metals and making oxygen, too. I wanted to make it by heating mercuric oxide—the way Priestley had first made it in 1774—

but I was afraid, until the fume cupboard was installed, of toxic mercury fumes. Yet it was easy to prepare simply by heating an oxygen-rich substance such as hydrogen peroxide or potassium permanganate. I remember thrusting a glowing wood chip into a test tube full of oxygen and seeing how it flared up, flamed with an intense brilliance.

I made other gases, too. I decomposed water, using electrolysis; and then I recomposed it, sparking hydrogen and oxygen together. There were many other ways of making hydrogen with acids or alkalis—with zinc and sulfuric acid or aluminum bottle caps and caustic soda. It seemed a shame to have this hydrogen just bubble off and go to waste, so to stopper my flasks, I got tight-fitting rubber bungs and corks, some with holes in the middle for glass tubes. One of the things I had learned in Uncle Dave's lab was how to soften glass tubing in a gas flame and gently bend it to an angle (and, more excitingly, to blow glass as well, gently puffing into the molten glass to make thin-walled globes and shapes of all sorts). Now, using glass tubing, I could light the hydrogen as it emerged from the stoppered flask. It had a colorless flame—not yellow and smoky like the flames of gas jets or the kitchen stove. Or I could feed the hydrogen, with a gracefully curved piece of glass tubing, into a soap solution to make soap bubbles filled with hydrogen; the bubbles, far lighter than air, would rush up to the ceiling and burst.

Occasionally I would collect hydrogen over water in an inverted trough. Holding the trough, still inverted, I could put this over my nose and breathe it in—it had no smell, no taste, there was no sensation whatever, but my voice would go all high and squeaky for a few seconds, a Mickey Mouse voice I could no longer recognize as my own.

I poured hydrochloric acid over chalk (though even a mild acid like vinegar would do), producing an effervescence of a different, much heavier gas, carbon dioxide. I could collect the heavy, invisible carbon dioxide in a beaker and see how a tiny balloon of air, much less dense, floated on it. Our fire extinguishers at home were filled with carbon dioxide, and these, too, I used occasionally for the gas.

When I filled a balloon with carbon dioxide, it sank to the floor heavily and stayed there—I wondered what it would be like to have a balloon filled with a really dense gas, xenon (five times denser than air). When I mentioned this to Uncle Tungsten, he told me of a tungsten compound, tungsten hexafluoride, which was nearly twelve times denser than air—it is the heaviest vapor we know of, he said. I had fantasies that one might find or make a gas as heavy as water, and bathe in it, float in it, as one floated in water. There was something about floating—floating and sinking—that continually exercised and energized me.[7]

I was mesmerized by the giant barrage balloons that floated overhead in wartime London, looking like vast aerial sunfish, with their plump, helium-filled bodies and trilobed tails. They were made of an aluminized fabric, so they gleamed brilliantly when the sun's rays hit them. They were attached to the ground by long cables, which (it was thought) could entangle enemy warplanes, prevent them from flying too low. The balloons were our giant protectors as well.

One such balloon was tethered in our cricket field, in Lymington Road, and this became the object of my special, ardent attentions. I would steal over from the cricket pitch when nobody was looking and touch the gently swelling, shining fabric softly; the balloons appeared only half-inflated on the ground, but when they reached their proper altitude in the air, the helium inside

---

[7]More than fifty years later (for my sixty-fifth birthday), I was able to gratify this boyhood fantasy, and had, besides the normal helium balloons, a few xenon balloons of astonishing density—as near to "lead balloons" as could be (tungsten hexafluoride, though denser, would have been too dangerous to use—it is hydrolyzed by moist air, producing hydrofluoric acid). If one twirled these xenon balloons in one's hand, then stopped, the heavy gas, by its own momentum, would continue rotating for a minute, almost as if it were a liquid.

them expanded, swelled them out fully. I loved the feel of the giant balloons, a feel which was doubtless half-erotic, although I did not realize this at the time. I often dreamed of the barrage balloons at night, imagining myself cradled, at peace, in their giant soft bodies, suspended, floating, far above the cluttered world, in a timeless empyrean ecstasy. Everyone, I think, was fond of the balloons—their straining upwards stood for optimism, made the heart beat faster—but for me the balloon in Lymington Road was special: it recognized and responded to my touch, I imagined, trembled (as I did) with a sort of rapture. It was not human, it was not animal, but it was in a sense animate; it was my first love object, the precursor, when I was ten.

11

HUMPHRY DAVY: A POET-CHEMIST

I first heard Humphry Davy's name, I think, a little before the war, when my mother took me to the Science Museum, up to the top floor, where there was a model of a coal mine, its dusty galleries lit by feeble lamps. There she showed me the Davy safety lamp—there were several models of this—and explained to me how it worked, and how it had saved innumerable lives. And then she showed me, next to it, the Landau lamp, invented in the 1870s by her father—basically an ingenious modification of the Davy one. Davy was thus identified in my mind as an ancestor of sorts, almost part of the family.

Born in 1778, Davy grew up at the beginning of Lavoisier's revolution. It was an age of discovery, the coming-of-age of chemistry—a time, too, when great theoretical clarifications were emerging. Davy, an artisan's son, was apprenticed to a local apothecary-surgeon in Penzance, but soon aspired to something larger. Chemistry, above all, started to attract him. He read and mastered Lavoisier's *Elements of Chemistry*—a remarkable achievement for an eighteen-year-old with little formal education. Grand (perhaps grandiose) visions started revolving in his mind: Could *he* be the new Lavoisier, perhaps the new Newton? (One of his notebooks from this time was labeled "Newton and Davy.")

Lavoisier had left a ghost of phlogiston in his conception of heat or "caloric" as an element, and in his first, seminal experiment, Davy melted ice by friction, thus showing that heat was motion, a form of energy, and not a material substance, as Lavoisier had thought. "The non-existence of caloric, or the fluid of heat, has been proved," Davy exulted. He set forth the results of his experiments in a long "Essay on Heat and Light," a critique of Lavoisier as well as a vision of a new chemistry that he hoped to found, one finally purged of all the remnants of alchemy and metaphysics.

When news of the young man, of his intellect and perhaps revolutionary new thoughts about matter and energy, reached the chemist Thomas Beddoes, he published Davy's essay, and invited him to his laboratory, the Pneumatic Institute in Bristol. Here Davy analyzed the oxides of nitrogen (which had first been isolated by Priestley)—nitrous oxide (N_2O), nitric oxide (NO), and the poisonous, brown "peroxide" of nitrogen (NO_2)—made a detailed comparison of their properties, and wrote a wonderful account of the effects of inhaling the fumes of nitrous oxide, "laughing gas." Davy's description of inhaling nitrous oxide, in its psychological perspicacity, is reminiscent of William James's own account of the same experience a century later, and it is perhaps the first description of a psychedelic experience in Western literature:

> A thrilling extending from the chest to the extremities was almost immediately produced . . . my visible impressions were

dazzling and apparently magnified, I heard distinctly every sound in the room. . . . As the pleasurable sensations increased, I lost all connection with external things; trains of vivid visible images passed through my mind and were connected with words in such a manner, as to produce perceptions perfectly novel. I existed in a world of newly connected and newly modified ideas. I theorised; I imagined that I made discoveries.

Davy also discovered that nitrous oxide was an anesthetic, and suggested its use in surgical operations. (He never followed up on this, and general anesthesia was only introduced in the 1840s, after his death.)

In 1800 Davy read Alessandro Volta's paper describing the first battery, his "pile"—a sandwich of two different metals with brine-dampened cardboard in between—which generated a steady electric current. Although static electricity, as lightning or sparks, had been explored in the previous century, no sustained electrical current was obtainable until now. Volta's paper, Davy was later to write, acted like an alarm bell among the experimenters of Europe, and, for Davy, suddenly gave form to what he now saw as his life's work.

He persuaded Beddoes to build a massive electric battery—it consisted of a hundred six-inch-square double plates of copper and zinc, and occupied an entire room—and started his first experiments with it a few months after Volta's paper. He suspected almost at once that the electric current was generated by chemical changes in the metal plates and wondered if the reverse was also true—whether one might induce chemical changes by the passage of an electric current.

Water could be created (as Cavendish had shown) by sparking hydrogen and oxygen together.[1] Could one now, with the new

---

[1] While Cavendish was the first to observe that hydrogen and oxygen, when exploded together, created water, he interpreted their reaction in terms of

power of electric current, do the opposite? In his very first electrochemical experiment, passing an electric current through water (he had to add a little acid to render it conducting), Davy showed that it could be decomposed into its constituent elements, hydrogen appearing at one pole or electrode of the battery, and oxygen at the other—though it was only several years later that he was able to show that they appeared in fixed and exact proportions.

---

phlogiston theory. Lavoisier, hearing of Cavendish's work, repeated the experiment, reinterpreting the results correctly, and claimed the discovery for himself, making no acknowledgment of Cavendish. Cavendish was unmoved by this, being wholly indifferent to matters of priority and, indeed, to all matters merely human or emotional.

While Boyle and Priestley and Davy were all eminently human and engaging, as well as scientifically brilliant, Cavendish was quite a different figure. The range of his achievements was astounding, from his discovery of hydrogen and his beautiful researches on heat and electricity to his famous (and remarkably accurate) weighing of the earth. No less astounding, and even in his lifetime the stuff of legend, was his virtual isolation (he rarely spoke to anyone, and insisted his servants communicate with him in writing), his indifference to fame and fortune (though he was the grandson of a duke, and for much of his life the richest man in England), and his ingenuousness and incomprehension in regard to all human relationships. I was deeply moved, but if anything more mystified, when I read more about him.

> He did not love; he did not hate; he did not hope; he did not fear; he did not worship as others do [wrote his biographer George Wilson in 1851]. He separated himself from his fellow men, and apparently from God. There was nothing earnest, enthusiastic, heroic, or chivalrous in his nature, and as little was there anything mean, grovelling, or ignoble. He was almost passionless. All that needed for its apprehension more than the pure intellect, or required the exercise of fancy, imagination, affection, or faith, was distasteful to Cavendish. An intellectual head thinking, a pair of wonderfully acute eyes observing, and a pair of very skilful hands experimenting or recording, are all that I realise in reading his memorials. His brain seems to have been but a calculating engine; his eyes inlets of vision, not fountains of tears; his hands instruments of manipulation which never trembled with emotion, or were clasped together in adoration, thanksgiving or despair; his heart only an anatomical organ, necessary for the circulation of the blood. . . .

With his battery, Davy found, he could not only electrolyze water, but heat metallic wires: a platinum wire, for example, could be heated to incandescence; and if the current was passed into rods of carbon, and these were then separated by a short distance, a dazzling electric "arc" would leap out and bridge them ("an arc so vivid," he wrote, "that even the sunlight compared with it appeared feeble"). Thus, almost casually, Davy hit upon what were to become two major forms of electrical illumination,

---

Yet, Wilson continued,

> Cavendish did not stand aloof from other men in a proud or supercilious spirit, refusing to count them his fellows. He felt himself separated from them by a great gulf, which neither they nor he could bridge over, and across which it was vain to stretch hands or exchange greetings. A sense of isolation from his brethren, made him shrink from their society and avoid their presence, but he did so as one conscious of an infirmity, not boasting of an excellence. He was like a deaf mute sitting apart from a circle, whose looks and gestures show that they are uttering and listening to music and eloquence, in producing or welcoming which he can be no sharer. Wisely, therefore, he dwelt apart, and bidding the world farewell, took the self-imposed vows of a Scientific Anchorite, and, like the Monks of old, shut himself up within his cell. It was a kingdom sufficient for him, and from its narrow window he saw as much of the Universe as he cared to see. It had a throne also, and from it he dispensed royal gifts to his brethren. He was one of the unthanked benefactors of his race, who was patiently teaching and serving mankind, whilst they were shrinking from his coldness, or mocking his peculiarities. . . . He was not a Poet, a Priest, or a Prophet, but only a cold, clear Intelligence, raying down pure white light, which brightened everything on which it felt, but warmed nothing—a Star of at least the second, if not of the first magnitude, in the Intellectual Firmament.

Many years later, I reread Wilson's astonishing biography and wondered what (in clinical terms) Cavendish "had." Newton's emotional singularities—his jealousy and suspiciousness, his intense enmities and rivalries—suggested a profound neurosis; but Cavendish's remoteness and ingenuousness were much more suggestive of autism or Asperger's syndrome. I now think Wilson's biography may be the fullest account we are ever likely to have of the life and mind of a unique autistic genius.

incandescence and arc lighting—though he did not develop these, but went on to other things.[2]

Lavoisier, making his list of elements in 1789, had included the "alkaline earths" (magnesia, lime, and baryta) because he felt they contained new elements—and to these Davy added the alkalis (soda and potash), for these, he suspected, contained new elements too. But there were as yet no chemical means sufficient to isolate them. Could the radically new power of electricity, Davy wondered, succeed here where ordinary chemistry had failed? First he attacked the alkalis, and early in 1807 performed the famous experiments that isolated metallic potassium and sodium by electric current. When this occurred, Davy was so ecstatic, his lab assistant recorded, that he danced with joy around the lab.[3]

One of my greatest delights was to repeat Davy's original experiments in my own lab, and I so identified with him that I could almost feel I was discovering these elements myself. Having read how he first discovered potassium, and how it reacted with water, I diced a little pellet of it (it cut like butter, and the cut surface glittered a brilliant silver-white—but only for an instant; it tarnished at once). I lowered it gently into a trough full of water and stood back—hardly fast enough, for the potassium caught fire instantly, melted, and as a frenzied molten blob rushed round and round in the trough, with a violet flame above

---

[2] The ease of obtaining hydrogen and oxygen by electrolysis, in ideally inflammable proportions, led at once to the invention of the oxy-hydrogen blowpipe, which produced higher temperatures than had ever been obtained before. This allowed, for example, the melting of platinum, and the raising of lime to a temperature at which it gave out the most brilliant sustained light ever seen.

[3] Mendeleev, sixty years later, was to speak of Davy's isolation of sodium and potassium as "one of the greatest discoveries in science"—great in its bringing a new and powerful approach to chemistry, in its defining of the essential qualities of a metal, and in its exhibition of the elements' twinship and analogy, the implication of a fundamental chemical group.

it, spitting and crackling loudly as it threw off incandescent fragments in all directions. In a few seconds the little globule had burned itself out, and tranquillity settled again over the water in the trough. But now the water felt warm, and soapy; it had become a solution of caustic potash, and being alkaline, it turned a piece of litmus paper blue.

Sodium was much cheaper and not quite as violent as potassium, so I decided to look at its action outdoors. I obtained a good-sized lump of it—about three pounds—and made an excursion to the Highgate Ponds in Hampstead Heath with my two closest friends, Eric and Jonathan. When we arrived, we climbed up a little bridge, and then I pulled the sodium out of its oil with tongs and flung it into the water beneath. It took fire instantly and sped around and around on the surface like a demented meteor, with a huge sheet of yellow flame above it. We all exulted—this was chemistry with a vengeance!

There were other members of the alkali metal family even more reactive than sodium and potassium, metals like rubidium and cesium (there was also the lightest and least reactive, lithium). It was fascinating to compare the reactions of all five by putting small lumps of each into water. One had to do this gingerly, with tongs, and to equip oneself and one's guests with goggles: lithium would move about the surface of the water sedately, reacting with it, emitting hydrogen, until it was all gone; a lump of sodium would move around the surface with an angry buzz, but would not catch fire if a small lump was used; potassium, in contrast, would catch fire the instant it hit the water, burning with a pale mauve flame and shooting globules of itself everywhere; rubidium was still more reactive, spluttering violently with a reddish violet flame; and cesium, I found, exploded when it hit the water, shattering its glass container. One never forgot the properties of the alkali metals after this.

Before Humphry Davy's discovery of sodium and potassium, metals were thought of as hard and dense and infusible, and here were ones as soft as butter, lighter than water, very easily melted, and with a chemical violence, an avidity to combine beyond anything ever seen. (Davy was so startled by the inflammability of sodium and potassium, and their ability to float on water, that he wondered whether there might not be deposits of these beneath the earth's crust, which, exploding upon the impact of water, were responsible for volcanic eruptions.) Could the alkali metals, indeed, be seen as true metals? Davy addressed this question just two months later:

> The great number of philosophical persons to whom this question has been put have answered in the affirmative. They agree with metals in opacity, lustre, malleability, conducting powers as to heat and electricity, and in their qualities of chemical combination.

After his success in isolating the first alkali metals, Davy turned to the alkaline earths and electrolyzed these, and within a few weeks he had isolated four more metallic elements — calcium, magnesium, strontium, and barium — all highly reactive and all able to burn, like the alkali metals, with brilliantly colored flames. These clearly formed another natural group.

Pure alkali metals do not exist in nature; nor do the elemental alkaline earth metals — they are too reactive and instantly combine with other elements.[4] What one finds instead are simple or complex salts of these elements. While salts tend to be nonconducting when crystalline, they can conduct an electric current well if dissolved in water or melted; and will indeed be decomposed by an electric current, yielding the metallic component of

---

[4] The enormous chemical reactivity of potassium made it a powerful new instrument in isolating other elements. Davy used it himself, only a year after he discovered it, to obtain the element boron from boric acid, and he tried to obtain silicon by the same method (Berzelius succeeded here, in 1824). Aluminum and beryllium, a few years later, were also isolated through the use of potassium.

the salt (e.g., sodium) at one pole, and the nonmetallic element (e.g., chlorine) at the other. This implied to Davy that the elements were contained in the salt as charged particles—why else should they be attracted to the electrodes? Why did sodium always go to one electrode and chlorine to the other? His pupil, Faraday, was later to call these charged particles of an element "ions," and further distinguished the positive and negative ones as "cations" and "anions." Sodium, in its charged state, was a strong cation, and chlorine, in its charged state, one of the strongest anions.

For Davy, electrolysis was a revelation that matter itself was not something inert, held together by "gravity," as Newton had thought, but was charged and held together by electrical forces. Chemical affinity and electrical force, he now speculated, were one and the same. For Newton and Boyle there had been only one force, universal gravity, holding not only the stars and planets together, but the very atoms of which they were composed. Now, for Davy, there was a second cosmic force, a force no less potent than gravity, but operating at the tiny distances between atoms, in the invisible, almost unimaginable, world of chemical atoms. Gravity, he felt, might be the secret of mass, but electricity was the secret of matter.

Davy loved to conduct experiments in public, and his famous lectures, or lecture-demonstrations, were exciting, eloquent, and often literally explosive. His lectures moved from the most intimate details of his experiments to speculation about the universe and about life, delivered in a style and with a richness of language that nobody else could match.[5] He soon became the most famous and influential lecturer in England, drawing huge crowds that

---

[5] Mary Shelley, as a child, was enthralled by Davy's inaugural lecture at the Royal Institution, and years later, in *Frankenstein,* she was to model Professor Waldman's lecture on chemistry rather closely on some of Davy's words when,

blocked the streets whenever he lectured. Even Coleridge, the greatest talker of his age, came to Davy's lectures, not only to fill his chemical notebooks, but "to renew my stock of metaphors." There still existed, in the early nineteenth century, a union of literary and scientific cultures—there was not the dissociation of sensibility that was so soon to come—and Davy's period at Bristol saw the start of a close friendship with Coleridge and the Romantic poets. Davy himself was writing (and sometimes publishing) a good deal of poetry at the time; his notebooks mix details of chemical experiments, poems, and philosophical reflections all together; and these did not seem to exist in separate compartments in his mind.[6]

There was an extraordinary appetite for science, especially chemistry, in these early, palmy days of the Industrial Revolution; it seemed a new and powerful (and not irreverent) way not only of

---

speaking of galvanic electricity, he said, "a new influence has been discovered, which has enabled man to produce from combinations of dead matter effects which were formerly occasioned only by animal organs."

[6] David Knight, in his brilliant biography of Davy, speaks of the passionate parallelism, the almost mystical sense of affinity and rapport, that Coleridge and Davy felt, and how the two planned, at one point, to set up a chemical laboratory together. In his book *The Friend,* Coleridge wrote:

> Water and flame, the diamond, the charcoal . . . are convoked and fraternized by the theory of the chemist. . . . It is the sense of a principle of connection given by the mind, and sanctioned by the correspondency of nature. . . . If in a *Shakespeare* we find nature idealized into poetry, through the creative power of a profound yet observant meditation, so through the meditative observation of a *Davy* . . . we find poetry, as it were, substantiated and realized in nature: yea, nature itself disclosed to us . . . as at once the poet and the poem!

Coleridge was not the only writer to "renew his stock of metaphors" with images from chemistry. The chemical term *elective affinities* was given an erotic connotation by Goethe; Keats, trained in medicine, reveled in chemical metaphors. Eliot, in "Tradition and the Individual Talent," employs chemical metaphors from beginning to end, culminating in a grand, Davyan metaphor for the poet's mind: "The analogy is that of the catalyst . . . The mind of the poet is the shred of platinum."

understanding the world but of moving it to a better state. Davy himself seemed to embody this new optimism, to be at the crest of a vast new wave of scientific and technological power, a power that promised, or threatened, to transform the world. He had discovered half a dozen elements, as a start, suggested new forms of lighting, made important innovations in agriculture, and developed an electrical theory of chemical combination, of matter, of the universe itself—all before the age of thirty.

In 1812, Davy, the son of a wood-carver, was knighted for his services to the empire—the first scientist so honored since Isaac Newton. In the same year he married, but this did not seem to distract him from his chemical researches in the least. When he set out for an extended honeymoon on the Continent, determined to do experiments and meet other chemists wherever he went, he brought along a good deal of chemical apparatus and various materials ("an airpump, an electrical machine, a voltaic battery . . . a blow-pipe apparatus, a bellows and forge, a mercurial and water gas apparatus, cups and basins of platinum and glass, and the common reagents of chemistry")—as well as his young research assistant, Michael Faraday. (Faraday, then in his early twenties, had followed Davy's lectures raptly, and wooed Davy by presenting him with a brilliantly transcribed and annotated version of them.)

In Paris, Davy had a visit from Ampère and Gay-Lussac, who brought with them, for his opinion, a sample of a shiny black substance obtained from seaweed, with the remarkable property that when heated, it did not melt, but turned at once into a vapor of a deep violet color. A year earlier, Davy had identified Scheele's greenish yellow "muriatic acid air" as a new element, chlorine. Now, with his enormous feeling for the concrete[7] and his genius

---

[7] The great chemist Justus von Liebig wrote powerfully about this feeling in his autobiography:

for analogy, Davy sensed that this odoriferous, volatile, highly reactive black solid might be another new element, an analog of chlorine, and soon confirmed that it was. He had already tried, unsuccessfully, to isolate Lavoisier's "fluoric radical," realizing that the element it contained, fluorine, would be a lighter and even more active analog of chlorine. But he also felt that the gap in physical and chemical properties between chlorine and iodine was so great as to suggest the existence of an intermediate element, as yet undiscovered, between them. (There was indeed such an element, bromine, but it fell not to Davy to discover it, but to a young French chemist, Balard, in 1826. Liebig himself, it turned out, had actually prepared the fuming brown liquid element before this, but misidentified it as "liquid iodine chloride"; after hearing of Balard's discovery, Liebig put the bottle in his "cupboard of mistakes.")

From France the wedding party moved by stages to Italy, with experiments along the way: collecting crystals from the rim of Vesuvius; analyzing gas from natural vents in the mountains (it turned out to be, Davy found, identical with marsh gas, or methane); and, for the first time, performing a chemical analysis

---

[Chemistry] developed in me the faculty, which is peculiar to chemists more than to other natural philosophers, of thinking in terms of phenomena; it is not very easy to give a clear idea of phenomena to anyone who cannot recall in his imagination a mental picture of what he sees and hears, like the poet and artist, for example. . . . There is in the chemist a form of thought by which all ideas become visible in the mind as the strains of an imagined piece of music. . . .

The faculty of thinking in phenomena can only be cultivated if the mind is constantly trained, and this was effected in my case by my endeavouring to perform, so far as my means would allow me, all the experiments whose description I read in the books . . . I repeated such experiments . . . a countless number of times, . . . till I knew thoroughly every aspect of the phenomenon which presented itself . . . a memory of the sense, that is to say of the sight, a clear perception of the resemblance or differences of things or of phenomena, which afterwards stood me in good stead.

of paint samples from old masterworks ("mere atoms," he announced).

In Florence, he experimented with burning a diamond under controlled conditions, with a giant magnifying glass. Despite the demonstration of diamond's inflammability by Lavoisier, Davy had been reluctant, up to this point, to believe that diamond and charcoal were, in fact, one and the same element. It was rather rare for elements to have a number of quite different physical forms (this was before the discovery of red phosphorus, or the allotropes of sulfur). Davy wondered if these might represent different forms of "aggregation" of the atoms themselves, but it was only much later, with the rise of structural chemistry, that this could be defined (the hardness of diamond, it was then shown, was due to the tetrahedral form of its atomic lattices, the softness and greasiness of graphite due to the packing of its hexagonal lattices in parallel sheets).

Davy returned to London after his honeymoon to one of the grandest practical challenges of his lifetime. The Industrial Revolution, now warming up, was devouring ever huger amounts of coal; coal mines were being dug ever deeper, deep enough now to run into the inflammable and poisonous gases of "fire-damp" (methane) and "choke-damp" (carbon dioxide). A canary carried down in a cage could serve as a warning of the presence of asphyxiating choke-damp; but the first indication of fire-damp was, all too often, a fatal explosion. It was desperately important to design a miner's lamp that could be carried into the lightless depths of the mines without any danger of igniting pockets of fire-damp.

Davy made a crucial observation—that a flame could not pass through a wire mesh or gauze, as long as this were kept cool.[8] He

---

[8] Davy went on with his investigations of flame, and, a year after the safety lamp, published *Some Philosophical Researches on Flame*. More than forty years

made many different sorts of lamps incorporating this principle, the simplest and most reliable being an oil lamp in which the only way air could get in or out was through screens of wire mesh. The perfected lamps were tried in 1816 and proved not only safe but also, by the appearance of the flame, reliable indicators of fire-damp.

In a further discovery, Davy found that if a platinum wire was put in an explosive mixture, it would become red-hot and glow. He had discovered the miracle of catalysis: how certain substances, such as the platinum metals, might induce a continuing chemical reaction on their surfaces, without being themselves consumed. Thus, for instance, the platinum loop we kept above the kitchen stove would glow when put in the stream of gas, and, becoming red-hot, ignite it. This principle of catalysis was to become indispensable in thousands of industrial processes.[9]

To an extent that I was only to realize later, Humphry Davy and his discoveries were part of our lives, from the electroplated cutlery to the catalytic gas-lighting loop, to photography (which he had been one of the first to perform, making photographs on leather, thirty years or more before others rediscovered the process), to the dazzling arc lamp used to project films in the local cinema. Aluminum, once costlier than gold (Napoleon III,

---

later, Faraday would return to the subject, in his famous Royal Institution lectures on *The Chemical History of a Candle.*

[9] Enlarging on Davy's observation of catalysis, Döbereiner found in 1822 that platinum, if finely divided, would not only become white-hot, but would ignite a stream of hydrogen passing over it. On this basis he made a lamp consisting basically of a tightly sealed bottle containing a piece of zinc which could be lowered into sulfuric acid, generating hydrogen. When the stopcock of the bottle was opened, hydrogen gushed out into a small container holding a bit of platinum sponge, and instantly burst into flame (a slightly dangerous flame, because it was virtually invisible, and one had to be cautious to avoid being burned). Within five years, there were twenty thousand Döbereiner lamps in use in Germany and England, so Davy had the satisfaction of seeing catalysis at work, indispensable in thousands of homes.

famously, used to give his guests plates of gold, while he himself dined on aluminum), had become cheap and available only with the use of Davyan electrolysis to extract it. And the thousand and one synthetics all around us, from artificial fertilizers to our gleaming bakelite telephones, were all made possible by the magic of catalysis. But, crucially, it was Davy's personality that appealed to me—not modest, like Scheele, not systematic, like Lavoisier, but filled with the exuberance and enthusiasm of a boy, with a wonderful adventurousness and sometimes dangerous impulsiveness—he was always at the point of going too far— and it was this which captured my imagination above all.

12

IMAGES

Photography had become another passion of mine, and my little lab, already so overfull, often doubled as a darkroom as well. If I try to remember what drew me to photography, I think of the chemicals involved—my hands were often stained with pyrogallol, and seemed to smell of "hypo," sodium hyposulfite, all the while; of the special lights—the deep ruby safelight; the large flashbulbs stuffed with shiny, crinkly, inflammable metal foil (usually magnesium or aluminum, occasionally zirconium). I think of the optics—the tiny, flattened image of the world on a ground-glass screen; the delight of different f-stops, of focusing, of different lenses; of all the intriguing emulsions one could use—it was, above all, the processes of photography that fascinated me.

But there was also, of course, the sense of being able to make a very personal and perhaps fugitive perception objective and permanent, especially as I lacked the ability to draw or paint. This was fanned, even before the war, by family photo albums, especially those which went back before my birth, to the beach scenes and bathing machines of the 1920s, the street scenes of London at the turn of the century, the stiffly posed grandparents and great-aunts and great-uncles of the 1870s. There were also, most precious of all, a couple of daguerreotypes, in special frames, dating from the 1850s; these had a detail, a finish, that seemed much finer, more brilliant, than those of the later paper prints. My mother particularly cherished one of these, a photo of her mother's mother, Judith Weiskopf, taken in Leipzig in 1853.

Then there was the whole wide world outside the family, the printed photos in books and newspapers, some of which struck me with great vividness, like the dramatic photos of the Crystal Palace on fire (these confirmed—or did they suggest?—my own very early memories of it) and photos of airships majestically floating (and another of a Zeppelin coming down in flames). I loved photos of distant people and places, most of all the photos in the *National Geographic* magazine, which would arrive, with its yellow-edged cover, each month. The *National Geographic,* moreover, had pictures in color, and these affected me especially. I had seen hand-tinted photos—Auntie Birdie was adept at such tinting—but I had never seen actual color photos before. An H. G. Wells tale, "The Queer Story of Brownlow's Newspaper," which I read around this time, describes how Brownlow receives one day, instead of his usual 1931 paper, a newspaper dated 1971. What first arouses Mr. Brownlow's attention, making him realize that he is confronting something incredible, is the fact that this newspaper has photographs in color—something inconceivable to him, living in the 1930s:

Never in all his life had he seen such colour printing—and the buildings and scenery and costumes in the pictures were strange. Strange and yet credible. They were colour photographs of actuality forty years from now.

I sometimes had such a feeling about the color pictures in the *National Geographic;* they, too, pointed to a brilliant, many-colored world of the future, and away from the monochromes of the past.

But I was more deeply drawn to the photographs of the past, with their dim, delicate sepia tones—they abounded in the older family albums and in the old magazines that I once found piled in the lumber room. I was already, by 1945, very conscious of change, of how prewar life had gone irredeemably, forever. But there were still photos, photos often casually taken, that now possessed a special value, photos of summer holidays before the war, photos of friends and neighbors and relatives, caught in the sunlight of 1935 or 1938, with no shadow or premonition of what would come. It seemed to me wonderful that photographs could capture actual moments, clean cross sections, as it were, of time, fixed forever in silver.

I longed to make photos myself, to document and chronicle scenes, objects, people, places, moments, before they changed or disappeared, swallowed by the transformations of memory and time. I took one such picture of Mapesbury Road, caught in the morning sunlight of July 9, 1945, my twelfth birthday. I wished to document, to hold forever, exactly what confronted me when I opened the curtains that morning. (I still have this photo, two photos, actually, designed to form a stereo pair, as a red and green anaglyph. Now, more than half a century later, it has almost replaced the actual memory, so that if I close my eyes and try to visualize the Mapesbury Road of my boyhood, all I see is the photograph I took.)

Such documentation was, in part, forced on me by the war, the wholesale way in which seemingly permanent objects were destroyed or removed. There had been wrought iron railings, beautiful and solid, around our front garden before the war, but when I returned home in 1943, they were no longer there. I found this very disturbing, and was even driven to doubt my own memory. Had there in fact been such railings before the war, or had I, in a fanciful or poetic way, somehow invented them? Seeing photos of my younger self, posed against the railings, was a great relief, proving that the railings were really there. And then there was the giant Cricklewood clock, the clock I remembered, or seemed to remember, at least, twenty feet high, with a golden face, in Chichele Road—this, too, was gone in '43. There was a similar clock in Willesden Green, and I assumed that I had somehow doubled it in my mind, endowing Cricklewood, my neighborhood, with its twin. Here again, it was a great relief, years later, to see a photo of this clock, to see that I had not invented it (both the iron railings and the clock had been removed as part of the war effort, when the country was desperate for all the iron it could obtain).

It was similar with the vanished Willesden Hippodrome, if indeed it had ever existed. If I even asked, I imagined, people would say, "Willesden Hippodrome, indeed! What's the boy thinking of? As if there would ever have been a *hippodrome* in Willesden!" It was only when I saw an old photo that my doubts were banished, and I became confident that there was indeed once such a hippodrome, though it was bombed out of existence during the war.

I read *1984* when it came out in 1949, and found its account of the "memory hole" peculiarly evocative and frightening, for it accorded with my own doubts about my memory. I think that reading this led to an increase in my own journal keeping, and photographing, and an increased need to look at testimonies of the past. This took many forms—an interest in antiquarian books and old things of every sort; in genealogy; in archaeology;

and most especially in paleontology. I had been introduced to fossils as a child by Auntie Len, but now I saw them as guarantors of reality.

So I loved old photos of our neighborhood and of London. They seemed to me like an extension of my own memory and identity, helped to moor me, anchor me in space and time, as an English boy born in the 1930s, born into a London similar to that in which my parents, my uncles and aunts, had grown up, a London which would have been recognizable to Wells, Chesterton, Dickens, or Conan Doyle. I pored over old photos, local and historical ones as well as the old family ones, to see where I came from, to see who I was.

If photography was a metaphor for perception and memory and identity, it was equally a model, a microcosm, of science at work—and a particularly sweet science, since it brought chemistry and optics and perception together into a single, indivisible unity. Snapping a picture, sending it out to be developed and printed, was exciting, of course, but in a limited way. I wanted to understand, to master for myself, all the processes involved, and to manipulate them in my own way.

I was especially fascinated by the early history of photography and the chemical discoveries that had led to it: how it was first realized, as early as 1725, that silver salts darkened with light, and how Humphry Davy (with his friend Thomas Wedgwood) had made contact images of leaves and insect wings on paper or white leather soaked in silver nitrate, and photos with a camera lucida. But they were unable to fix the images they produced and could view them only in red light or candlelight, otherwise they would blacken completely. I wondered why Davy, so expert a chemist and so familiar with Scheele's work, had failed to make use of Scheele's observation that ammonia could "fix" the images (by removing the surplus silver salt)—had he done so, he might have been seen as the father of photography, anticipating the final breakthrough in the 1830s, when Fox Talbot, Daguerre, and others were able to make permanent images, using chemicals to develop and fix them.

We lived very near my cousin Walter Alexander (it was to his flat we went when a bomb landed next door during the Blitz), and I became close to him despite the great disparity in our ages (though my first cousin, he was thirty years my senior), for he was a professional magician and photographer who retained a very playful character all his life, and loved tricks and illusions of every sort. It was Walter who first inducted me into photography, by showing me the magic of an image emerging as he developed sheets of film in his red-lit darkroom. I never tired of the wonder of this, seeing the first faint hints of an image—were they really there, or was one deceiving oneself?—grow stronger, richer, clearer, come to full life, as he tilted the film to and fro in the tray of developing fluid, until at last, fully developed, there lay a tiny, perfect facsimile of the scene.

Walter's mother, Rose Landau, had gone to South Africa with her brothers in the 1870s, where she took photographs of mines and miners, taverns and boomtowns, in the early days of the diamond and gold rushes. It had required considerable physical strength, as well as audacity, to make such photographs at this time, for she had to lug a massive camera around with her, along with all the glass plates it might need. Rose was still alive in 1940, the only one of the firstborn uncles and aunts I ever met. Walter himself had her original camera, as well as a considerable collection of cameras and stereoscopes of his own.

In addition to an original Daguerre camera, complete with its iodizing and mercury boxes, Walter had a huge view camera, with a rising front and tilt and bellows, that took eight-by-ten-inch sheet film (he still used this, at times, for studio portraits); a stereo camera; and a beautiful little Leica, with an f/3.5 lens—the first 35-millimeter miniature camera I had seen. The Leica was his favorite camera when he went hiking; he preferred to use a twin-lens reflex, a Rolleiflex, for general use. He also had some trick cameras from the beginning of the century—one of these, built for detective work, looked just like a pocket watch, and took pictures on 16-millimeter film.

All my own photography at first was in black and white—I could not have developed and printed my own films, otherwise—but I had no sense that these were "lacking" color. My first camera was a pinhole camera, which gave surprisingly good pictures, with an enormous depth of focus. Then I had a simple fixed-lens box camera—it cost two shillings at Woolworth's. Then a folding Kodak camera, which took 620 roll film. I was fascinated by the speeds and finenesses of different emulsions, from the slow, fine-grained ones which allowed exquisite detail to the fastest ones, almost fifty times faster than some of the slow emulsions, so that one could take photographs even at night (though these were so grainy one could scarcely enlarge them at all). I looked at some of these different emulsions under the microscope, seeing what the grains of silver actually looked like, and wondered whether one could have grains of silver so small as to produce a virtually grainless emulsion.

I enjoyed making light-sensitive emulsions myself, absurdly crude and slow as they were compared to the ready-made ones. I would take a 10-percent solution of silver nitrate and add it slowly, with continual stirring, to a solution of potassium chloride and gelatin. The crystals suspended in the gelatin were extremely fine and not too light-sensitive, so one could do this safely under a red light. One could make the crystals larger and more sensitive by warming the emulsion for several hours, which would allow the smallest crystals to redissolve and redeposit on the larger ones. After this "ripening," one added a little more gelatin, let it all set to a stiff jelly, and then smeared it on paper.

I could also impregnate paper directly with silver chloride, avoiding the gelatin altogether, by first immersing the paper in a salt solution and then in silver nitrate; the silver chloride formed would be held by the fibers of the paper. Either way, I was able to make my own print-out paper, as it was called, and with this to make contact prints from negatives, or silhouettes of lace or ferns, though it took several minutes of exposure to direct sunlight to obtain these.

Fixing the prints with hypo straight after exposure tended to produce rather ugly brown colors, and this drew me into experimenting with toning of various sorts. The simplest was sepia toning—not (alas) done with cuttlefish ink, sepia, as I had hoped, but by converting the silver of the image to sepia-colored silver sulfide. One could do gold toning, too—this involved immersion in a solution of gold chloride, and produced a bluish purple image, metallic gold being precipitated onto the particles of silver. And if one tried this after sulfide toning, one could get a lovely red color, an image of gold sulfide.

I soon spread from this to other forms of toning. Selenium toning gave a rich reddish color, and palladium- and platinum-toned prints had a fine, sober quality, more delicate, it seemed to me, than the usual silver prints. One had to start with a silver image, of course, because only silver salts were sensitive to light, but then one could replace it with almost any other metal. One could easily replace the silver with copper, uranium, or vanadium. A particularly wild combination was to combine a vanadium salt with an iron salt such as ferric oxalate, and then the yellow of the vanadium ferrocyanide and the blue of the ferriferrocyanide would combine to form a brilliant green. I enjoyed disconcerting my parents with pictures of green sunsets, green faces, and fire engines or double-decker buses turned green. My photographic manual also described toning with tin, cobalt, nickel, lead, cadmium, tellurium, and molybdenum—but I had to stop myself at this point, for I was becoming obsessed, going overboard with toning, with the possibility of pressing all the metals I knew into use in the darkroom, and forgetting what photography was really for. This sort of too-muchness had no doubt been noticed at school, for it was around this time that I received a school report that said, "Sacks will go far, if he does not go too far."

There was an oddly massive, chunky camera in Walter's collection—this, he said, was a color camera: it had two half-silvered mirrors in it, dividing the incoming light into three beams, and these were directed through differently colored filters to three separate plates. Walter's color camera was a direct descendant of a famous experiment done at the Royal Institution by Clerk Maxwell in 1861, photographing a colored bow with ordinary black-and-white plates through filters of the three primary colors—red, green, and violet—and projecting the black-and-white positives of these images using three lanterns with corresponding filters. When they were all perfectly superimposed, the three black-and-white pictures exploded into full color. With this, Maxwell showed that every color visible to the human eye could be constructed from just these three "primary" colors, because the eye itself had three equivalently "tuned" color receptors, rather than an infinity of color receptors for every conceivable hue and wavelength.

While Walter once demonstrated this to me with three lanterns, I was eager to have this miracle, this sudden explosion of color, more immediately to hand. The most exciting way of getting instant color was by a process called Finlaycolor, in which, in effect, three color-separation negatives were taken simultaneously by using a grid ruled with microscopic red, green, and violet lines. One then made a positive, a lantern slide, from this negative, and brought it into exact alignment with the grid. This was tricky, delicate, but when one had them in perfect register, the previously black-and-white slide would burst into full color. Since the screen, with its microscopic lines, simply appeared grey, one saw, when it was juxtaposed with the slide, the most magical, unexpected creation of color, where seemingly there had been none before. (The *National Geographic* originally used Finlaycolor, and one could see the fine lines on these if one looked with a magnifying glass.)

To make color prints, one had to print three positive images in the complementary colors—cyan, magenta, and yellow—and then superimpose them. Though there was a film, Kodachrome, that did this automatically, I preferred to do it in the old, delectable way, making separate cyan, magenta, and yellow diapositives from my separation negatives and then floating them gently, one above the other, until I had them in exact superposition. With this, suddenly, marvelously, the colors of the original burst out, having been coded, as it were, in the three monochromes.

I fiddled with these color separations endlessly, seeing the effect of juxtaposing two rather than three colors, or viewing the slides through the wrong filters. These experiments were at once amusing and instructive; they allowed me to create a range of strange color distortions, but above all they taught me to admire the elegance and economy with which the eye and brain worked, and which one could simulate remarkably well with a photographic three-color process.

We also had at the house hundreds of stereoscopic "views"— many on cardboard rectangles, others on glass plates—paired, faded sepia photos of Alpine scenery, the Eiffel Tower, Munich in the 1870s (my mother's mother was born in Gunzenhausen, a little village some miles from Munich), Victorian beach and street scenes, and industrial scenes of various sorts (one particularly arresting view was of a Victorian factory, with long treadles driven by steam engines, and it was this image that came to my mind when I read about Coketown in *Hard Times*). I loved feeding these double photographs into the big stereoscope in the drawing room—a massive wooden instrument that stood on its own stand and had brass knobs for focusing and altering the separation of the lenses. Such stereoscopes were still quite common, though no longer as universal as they had been at the turn of the century. Seeing the flat, dim photographs suddenly acquire a new dimension, a real and intensely visible depth, gave them a special reality, a verisimilitude of a peculiar and private sort. There was a romantic, secret quality to the stereo views, for one was privy to a sort of frozen theater when one looked through the eyepieces— a theater entirely one's own. I felt I could almost enter into them, like the dioramas in the museum.

There was, in these views, a small but crucial difference of parallax or perspective between the two pictures, and it was this which created the sense of depth. One had no sense of what each eye saw separately, for the two views coalesced, magically, to form a single coherent view.

The fact that depth was a construction, a "fiction" of the brain, meant that one could have deceptions, illusions, tricks of various sorts. I never had a stereo camera myself, but would take two pictures in succession, moving the camera a couple of inches between exposures. If one moved the camera more than this, the parallactic differences were exaggerated, and the two pictures, when fused, gave an exaggerated sense of depth. I made a hyper-stereoscope, using a cardboard tube with mirrors set obliquely inside it, increasing the interocular distance, in effect, to two feet or more. This was marvelous for bringing out the different depths of distant buildings or hills, but yielded bizarre effects at close distances—a Pinocchio effect, for example, when one looked at people's faces, for their noses seemed to be sticking out inches in front of them.

It was also intriguing to reverse the pictures. One could easily do this with stereo photographs, but one could also do it by making a pseudoscope, with a short cardboard tube and mirrors, so that the apparent position of the eyes was reversed. This caused distant objects to look closer than nearby ones—a face, for instance, might look like a concave mask. But it produced an interesting rivalry or contradiction, for one's knowledge, and every other visual cue, might be saying one thing, and the pseudoscopic images saying another, and one would see first one thing then another, as the brain alternated between different perceptual hypotheses.[1]

The other side to all of this, I came to realize—a sort of deconstruction or decomposition—could occur when I had migraines, in which there were often strange visual alterations. My sense of color might be briefly lost or altered; objects might look flat, like cutouts; or instead of seeing movement normally, I might see a series of flickering stills, as when Walter ran his film projector too slow. I might lose half of my visual field, with objects missing to one side, or faces bisected. I was terrified when I first got attacks like this—they started when I was four or five, before the war—but when I told my mother about them, she said she had similar attacks, and that they did no harm and lasted only a few minutes. With this, I started to look forward to my occasional attacks, wondering what might happen in the next one (no two were quite the same), what the brain, in its ingenuity, might be up to. Migraines and photography, between them, may have helped to tilt me in the direction in which, years later, I would go.

---

[1] I was intrigued, too (though I never practiced it), by cine-photography. Here again it was Walter who made me realize that there was no actual movement in the film, only a succession of still images which the brain synthesized to give an impression of movement. He demonstrated this to me with his film projector, slowing it down to show me only the still images, and then speeding it up until the illusion of motion suddenly occurred. He had a zoetrope, with images painted on the inside of a wheel, and a thaumatrope, with drawings on a stack of cards, which when rotated, or rapidly flicked, would give the same illusion. So I had the sense that movement, too, was constructed by the brain, in a manner analogous to that of color and depth.

My brother Michael was fond of H. G. Wells, and lent me his copy of *The First Men in the Moon* at Braefield. It was a small book, bound in blue morocco leather, and its illustrations impressed me as much as the text—the attenuated Selenites, walking in single file, and the Grand Lunar, with his distended brain case, in his luminous, fungus-lit cavern on the moon. I loved the optimism and excitement of the journey to space, and the idea of a material ("cavorite") impermeable to gravity. One of the chapters was called "Mr. Bedford in Infinite Space," and I loved the notion of Mr. Bedford and Mr. Cavor in their little sphere (it resembled Beebe's bathysphere, which I had seen pictures of), snapping the cavorite shutters open and closed, shutting off the earth's gravity. The Selenites, the moon people, were the first aliens I had ever read about, and after this I sometimes met them in my dreams. But there was sadness, too, because Cavor in the end is marooned on the moon, with only the inhuman, insectile Selenites for company, in unutterable loneliness and solitude.

After Braefield, *The War of the Worlds* became a favorite too, not least because the Martian fighting machines generated an exceedingly dense, inky vapor ("it sank down through the air and poured over the ground in a manner rather liquid than gaseous") that contained an unknown element, combined with the gas argon—and I knew that argon, an inert gas, could not be compounded by any earthly means.[2]

I was very fond of bicycling, especially on country roads through the little towns and villages around London, and reading *The War of the Worlds*, I decided to trace the advance of the Martians, starting on Horsell Common, where the first Martian cylinder landed. Wells's descriptions seemed so real to me that by

---

[2] Wells's reference to the Martians' unknown element also intrigued me later when I learned about spectra, for he described it, early in the book, as "giving a group of four lines in the blue of the spectrum," though subsequently—did he reread what he had written?—as giving "a brilliant group of three lines in the green."

the time I reached Woking, I found it surprisingly intact, considering how it had been devastated by the Martian heat ray in '98. And I was startled, in the little village of Shepperton, at finding the church steeple still standing, for I had accepted, almost as historical fact, that it had been knocked down by a reeling Martian tripod. And I could not go to the Natural History Museum without thinking of "the magnificent and almost complete specimen [of a Martian] in spirits" which Wells assured us was there. (I would find myself looking for this in the cephalopod gallery, as all the Martians seemed to be somewhat octopoid in nature.)

It was similar with the Natural History Museum itself—its ruined, cobwebbed galleries open to the air—which Wells's Time Traveller wanders through in A.D. 800,000. I could never go to the museum thereafter without seeing its desolate future form superimposed on the present, like the memory of a dream. Indeed, the pedestrian reality of London itself became transformed for me by the charged and mythical London of Wells's short stories, with places that could only be seen in certain moods or states—the door in the wall, the magic shop.

I found the later, "social" Wells novels of little interest as a boy, preferring the earlier tales, which combined remarkable science-fiction extrapolations with an intense, poetic sense of human frailty and mortality, as with the Invisible Man, so arrogant at first, who dies so pitifully, or the Faustian Dr. Moreau, who is finally killed by his own creations.

But his stories were also full of ordinary people who have extraordinary visual experiences of every sort: the little shopkeeper who is granted ecstatic visions of Mars through gazing into a mysterious crystal egg; or the young man whose eyes are given a sudden twist as he stands between the poles of an electromagnet in a storm, transporting him visually to an uninhabited rock near the South Pole. I was addicted to Wells's stories, his fables, as a boy (and many are still resonant for me fifty years later). The fact that he was still alive in 1946, still with us, after the war, made me long urgently, improperly, to see him. And having heard that he lived in a little terrace of houses, Hanover Terrace, off Regent's Park, I would sometimes go there after school, or on weekends, hoping to catch a glimpse of the old man.

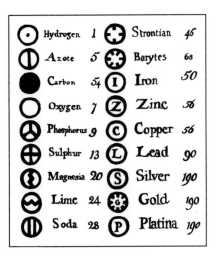

⊙	Hydrogen	1	✦ Strontian	46
⏀	Azote	5	✸ Barytes	68
●	Carbon	54	Ⓘ Iron	50
○	Oxygen	7	Ⓩ Zinc	56
⊛	Phosphorus	9	Ⓒ Copper	56
⊕	Sulphur	13	Ⓛ Lead	90
⊗	Magnesia	20	Ⓢ Silver	190
⊗	Lime	24	✹ Gold	190
⏀	Soda	28	Ⓟ Platina	190

MR. DALTON'S ROUND BITS OF WOOD

Experimenting in my lab brought home to me that chemical mixtures were completely unlike chemical compounds. One could mix salt and sugar, say, in any proportion. One could mix salt and water—the salt would dissolve, but then one could evaporate it and recover the salt unchanged. Or one could take a brass alloy and recover its copper and zinc unchanged. When one of my dental fillings came out, I was able to distill off its mercury, unchanged. All of these—solutions, alloys, amalgams—were mixtures. Mixtures, basically, had the properties of their ingredients (plus one or two "special" qualities perhaps—the relative hardness of brass, for example, or the lowered freez-

ing point of salt water). But compounds had utterly new properties of their own.

It was tacitly accepted by most chemists in the eighteenth century that compounds had fixed compositions and the elements
in them would combine in precise, invariable proportions—
practical chemistry could hardly have proceeded otherwise. But
there had been no explicit investigations of this, or declarations
on the matter, until Joseph-Louis Proust, a French chemist working in Spain, embarked on a series of meticulous analyses com-
paring various oxides and sulfides from around the world. He was
soon convinced that all genuine chemical compounds did indeed
have fixed compositions—and that this was so however the compound was made, or wherever it was found. Red mercuric sulfide,
for instance, always had the same proportions of mercury and sulfur, whether it was made in the lab or found as a mineral.[1]

---

[1] Yet Proust's view was challenged by Claude-Louis Berthollet. A senior
chemist of great eminence, an ardent supporter of Lavoisier (and a collaborator
with him on the Nomenclature), Berthollet had discovered chemical bleaching
and accompanied Napoleon as a scientist on his 1798 expedition to Egypt. He
had observed that various alloys and glasses manifestly had quite varied chemical compositions; therefore, he maintained, compounds could have a continuously variable composition. He also remarked, when roasting lead in his
laboratory, a striking, continuous color change—did this not imply a continuous absorption of oxygen with an infinite number of stages? It was true, Proust
argued, that heated lead took up oxygen continuously and changed color as it
did so, but this was due, he thought, to the formation of three distinctly colored oxides: a yellow monoxide, then red lead, then a chocolate-colored dioxide—admixed like paints, in varying proportions, depending on the state of
oxidation. The oxides themselves might be mixed together in any proportion,
he felt, but each was itself of fixed composition.

Berthollet also wondered about such compounds as ferrous sulfide, which
never contained exactly the same proportions of iron and sulfur. Proust was
unable to give a clear answer here (and indeed the answer only became clear
with a subsequent understanding of crystal lattices and their defects and sub-
stitutions—thus sulfur can substitute for iron in the iron sulfide lattice to a
variable extent, so that its effective formula varies from Fe_7S_8 to Fe_8S_9. Such
nonstoichiometric compounds came to be called berthollides).

Between pole and pole [Proust wrote] compounds are identical in composition. Their appearance may vary owing to their mode of aggregation, but their properties never. . . . The cinnabar of Japan has the same composition as the cinnabar of Spain; silver chloride is identically the same whether obtained from Peru or from Siberia; in all the world there is but one sodium chloride; one saltpetre; one calcium sulfate; and one barium sulfate. Analysis confirms these facts at every step.

By 1799, Proust had generalized his theory into a law—the law of fixed proportions. Proust's analyses, and his mysterious law, excited attention among chemists everywhere, not least in England, where they were to inspire profound insights in the mind of John Dalton, a modest Quaker schoolteacher in Manchester.

Gifted in mathematics, and drawn to Newton and his "corpuscular philosophy" from an early age, Dalton had sought to understand the physical properties of gases—the pressures they exerted, their diffusion and solution—in corpuscular or "atomic" terms. Thus he was already thinking of "ultimate particles" and their weights, albeit in this purely physical context, when he first heard of Proust's work, and by a sudden intuitive leap, saw how these ultimate particles might account for Proust's law, and indeed the whole of chemistry.

For Newton and Boyle, though there were different forms of matter, the corpuscles or atoms of which they were composed were all identical. (Thus there was always, for them, the alchemical possibility of turning a base metal into gold, for this only entailed change of form, a transformation of the same basic mat-

---

Thus both Proust and Berthollet were right in a way, but the vast majority of compounds were Proustian, with a fixed composition. (And it was perhaps necessary that Proust's view became the favored one, for it was Proust's law which was to inspire the profound insights of Dalton.)

ter.)[2] But now the concept of elements, thanks to Lavoisier, was clear, and for Dalton there were as many kinds of atoms as there were elements. Every one had a fixed and characteristic "atomic weight," and this was what determined the relative proportions in which it combined with other elements. Thus if 23 grams of sodium invariably combined with 35.5 grams of chlorine, this was because sodium and chlorine atoms had atomic weights of 23 and 35.5. (These atomic weights were not, of course, the actual weights of atoms, but their weights relative to that of a standard—for example, that of a hydrogen atom.)

Reading Dalton, reading about atoms, put me in a sort of rapture, thinking that the mysterious proportionalities and numbers one saw on a gross scale in the lab might reflect an invisible, infinitesimal, inner world of atoms, dancing, touching, attracting, and combining. I had the sense that I was being enabled to see, using the imagination as a microscope, a tiny world, an ultimate world, billions or trillions of times smaller than our own—the actual constituents of matter.

Uncle Dave had shown me gold leaf, beaten and hammered out until it became almost transparent, so that it transmitted light, a beautiful bluish green light. This leaf, a millionth of an inch thick, he said, was only a few dozen atoms thick. My father had shown me how a very bitter substance such as strychnine could be diluted a millionfold and still be tasted. And I liked to experiment with thin films, to blow soap bubbles in the bath— a speck of soapy water could be blown, with care, into a huge bubble—and to watch oil, in iridescent films, spreading on wet roads. All these prepared me, in a way, to imagine the very

---

[2] Though Newton hinted, in his final *Quaerie,* at something that almost seems to prefigure a Daltonian concept:

> God is able to create particles of matter of several sizes and figures, and in several proportions to the space they occupy, and perhaps of different densities and forces.

My grandparents, Marcus and Chaya Landau, in 1902 with their thirteen children, in the garden of their house in Highbury New Park. Standing: Mick, Violet, Isaac, Abe, Dora, Sydney, Annie. Seated: Dave, Elsie (my mother), Len, my grandfather and grandmother, Birdie. In front: Joe and Doogie.

In my mother's arms in late 1933, with
Marcus, David, Pop, and Michael.

As a boy of three, before the war.

Together on our last boat ride before
the war, in August 1939, near Bournemouth: David, Pop, Michael, Ma, me,
and Marcus.

A brief visit home from Braefield in the winter of 1940:
Pop, me, Ma, Michael, and
David (Marcus, the eldest, was
already away at university).

Among my fellow wolf-cubs at
The Hall, 1943.

My bar mitzvah photo, in
front of the house, 1946.

Uncle Dave (left)
and Uncle Abe,
photographed at a
Tungstalite company
outing in 1938.

Auntie Birdie.

Auntie Len, at
Delamere after
the war.

small—the smallness of particles that composed the millionth-of-an-inch thickness of gold leaf, a soap bubble, or an oil film.

But what Dalton intimated was infinitely more thrilling: for it was not just atoms in the Newtonian sense, but atoms as richly individual as the elements themselves—atoms whose individuality *gave* elements theirs.

Dalton later made wooden models of atoms, and I saw his actual models in the Science Museum as a boy. These, crude and diagrammatic as they were, excited my imagination, helped give me a sense that atoms really existed. But not everyone felt this, and, for some chemists, Dalton's models epitomized the absurdity, as they saw it, of an atomic hypothesis. "Atoms," the eminent chemist H. E. Roscoe was to write, eighty years later, "are round bits of wood invented by Mr. Dalton."

It was indeed possible, in Dalton's time, to regard the idea of atoms as implausible, if not outright nonsense, and it would be over a century before indisputable evidence for the existence of atoms was secured. Wilhelm Ostwald, for one, was not convinced of the reality of atoms, and in his 1902 *Principles of Inorganic Chemistry* he wrote:

> Chemical processes occur in such a way as if the substances were composed of atoms. . . . At best there follows from this the *possibility* that they are in reality so: not however, the *certainty*. . . . One must not be led astray by the agreement between picture and reality, and confound the two. . . . An hypothesis is only an *aid to representation.*

Now, of course, we can "see" and even manipulate individual atoms, using an atomic force microscope. But it required enormous vision and courage, at the very beginning of the nineteenth century, to postulate entities so utterly beyond the bounds of any empirical demonstration possible at the time.[3]

---

[3] Dalton represented the atoms of elements as circles with internal designs, sometimes reminiscent of the symbols of alchemy, or the planets; while the

Dalton's theory of chemical atoms was detailed in his notebook on the 6th of September, 1803, his thirty-seventh birthday.
He was at first too modest or too diffident to publish anything on
his theory (he had, however, worked out the atomic weights of
half a dozen elements—hydrogen, nitrogen, carbon, oxygen,
phosphorus, sulfur—which he recorded in his notebook). But
word was soon out that he had hatched something astonishing,
and Thomas Thomson, the eminent chemist, went up to Manchester to meet him. A single short conversation with Dalton in
1804 "converted" Thomson, altered his life. "I was enchanted,"
he later wrote, "with the new light which immediately burst
upon my mind, and I saw at a glance the immense importance of
such a theory."

Although Dalton had presented some of his thoughts to the
Literary and Philosophical Society in Manchester, they did not
become known to a wider public until Thomson wrote of them.
Thomson's presentation was brilliant and persuasive, much more
so than Dalton's own exposition, which was crammed, awkwardly, into the final pages of his 1808 *New System*.

But Dalton himself realized that there were fundamental
problems with his theory. For to pass from a combining or equivalent weight to an atomic weight required that one know the
exact formula of a compound, for the same elements, in some
cases, might combine in more than one way (as in the three
oxides of nitrogen). So Dalton assumed that if two elements

---

compound atoms (which we would now call "molecules") had increasingly
intricate geometric configurations—the first premonition of a structural
chemistry that was not to be developed for another fifty years.

Though Dalton spoke of his atomic "hypothesis," he was convinced that
atoms really existed—hence his violent objection to the terminology Berzelius
was to introduce, in which an element was denoted by one or two letters of its
name rather than his own iconic symbol. Dalton's passionate opposition to
Berzelius's symbolism (which he felt concealed the actuality of atoms) lasted to
the end of his life, and indeed when he died in 1844 it was from a sudden
apoplexy, following a violent argument defending the realness of his atoms.

formed only a single compound (as hydrogen and oxygen appeared to do in water, or nitrogen and hydrogen in ammonia), they would do so in the simplest possible ratio: one to one. This ratio, he felt, would surely be the most stable. Thus he took the formula of water (in modern nomenclature) to be HO, and the atomic weight of oxygen to be the same as its equivalent weight, namely 8. Similarly, he took the formula of ammonia to be NH, and thus the atomic weight of nitrogen to be 5.

And yet, as was demonstrated by the French chemist Gay-Lussac, in the very year that Dalton published his *New System,* if one measured volumes and not weights one found that *two* volumes, not one, of hydrogen combined with one volume of oxygen, to yield two volumes of steam. Dalton was skeptical of these findings (although he could have confirmed them himself with great ease), skeptical because he felt they would entail the breaking of an atom into two, to allow the combination of a half-atom of oxygen with each atom of hydrogen.

Although Dalton talked about "compound" atoms, he had not clearly distinguished (any more clearly than his predecessors) between molecules—the smallest amount of an element or compound that could exist free—and atoms—the actual units of chemical combination. The Italian chemist Avogadro, reviewing Gay-Lussac's results, now hypothesized that equal volumes of gases contained equal numbers of *molecules.* For this to be so, the molecules of hydrogen and oxygen would have to have two atoms apiece. Their combination to form water, therefore, could be represented as $2H_2 + 1O_2 \rightarrow 2H_2O$.

But in an extraordinary way (at least so it seems in retrospect), Avogadro's suggestion of diatomic molecules was ignored or resisted by virtually everyone, including Dalton. There remained great confusion between atoms and molecules, and a disbelief that atoms of the same sort could link together. There was no problem in seeing water, a compound, as H_2O, but a seemingly insuperable difficulty in allowing that a molecule of pure hydrogen could be H_2. Many atomic weights of the early nineteenth century were thus wrong by simple numerical factors—some seemed to be half what they should be, some twice, some a third, some a quarter, and so on.

Griffin's book, my first guide in the laboratory, was written in the first half of the nineteenth century, and many of his formulas, and hence many of his atomic weights, were as erroneous as Dalton's. Not that any of this mattered too much in practice—nor, indeed, did it affect the great virtue, the many virtues, of Griffin. His formulas and his atomic weights might indeed have been wrong, but the reagents he suggested, and their quantities, were exactly right. It was only the interpretation, the formal interpretation, that was askew.

With such confusion about elemental molecules, added to uncertainty about the formulas of many compounds, the very notion of atomic weights started to be discredited in the 1830s, and indeed the very notion of atoms and atomic weights fell into disrepute, so much so that Dumas, the great French chemist, exclaimed in 1837, "If I were master I would efface the word atom from science."

Finally in 1858, Avogadro's countryman Stanislao Cannizzaro realized that Avogadro's 1811 hypothesis provided an elegant way out of the decades-long confusion about atoms and molecules, atomic and equivalent weights. Cannizzaro's first paper was as ignored as Avogadro's had been, but when, at the close of 1860, chemists gathered at the first-ever international chemical meeting in Karlsruhe, it was Cannizzaro's presentation that stole the show, and ended the intellectual agony of many years.

This was some of the history I nosed out when I emerged from my lab and got a ticket to the library of the Science Museum in 1945. It was evident that the history of science was anything but a straight and logical series, that it leapt about, split, converged, diverged, took off at tangents, repeated itself, got into jams and corners. There were some thinkers who paid little attention to history (and it may be that there are many original workers who are much better off for not knowing their precursors or antecedents—Dalton, one feels, might have had more difficulty in proposing his atomic theory had he known the huge and confused history of atomism for the two thousand years that preceded him). But there were others who pondered the history of their subjects continually, and whose own contributions were integrally related to their pondering—and it is clear that this was the case with Cannizzaro. Cannizzaro thought intensely about Avogadro; saw the implications of his hypothesis as no one else had; and with them, and his own creativity, revolutionized chemistry.

Cannizzaro felt very passionately that the history of chemistry needed to be in the minds of his students. In a beautiful essay on the teaching of chemistry, he described how he introduced his pupils to its study by "endeavouring to place them . . . on the same level with the contemporaries of Lavoisier," so that they might experience, as Lavoisier's contemporaries did, the full revolutionary force, the wonder of his thought; and then a few years ahead, so that they could experience the sudden, blinding illumination of Dalton.

"It often happens," Cannizzaro concluded, "that the mind of a person who is learning a new science, has to pass through all the phases which the science itself has exhibited in its historical evolution." Cannizarro's words had a powerful resonance for me, because I, too, in a way, was living through, recapitulating, the history of chemistry in myself, rediscovering all the phases through which it had passed.

14

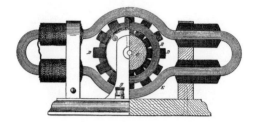

LINES OF FORCE

When I was very young I had been intrigued by "frictional" electricity, of the sort that made rubbed amber attract bits of paper, and when I returned from Braefield, I began to read about "electrical machines"—discs or globes of some nonconducting material, turned by a crank and rubbed against the hand, or a cloth, or a cushion of some sort—which would produce powerful sparks or shocks of static electricity. It seemed easy enough to make such a simple machine, and in my first attempt at making one I used an old record as the disc. Gramophone records at the time were made of vulcanite and easily electrified; the only problem was that they were thin and fragile, easily shattered. For a second, more robust machine, I used a thick glass plate and a cushion covered with leather and coated with zinc amalgam. I could get handsome sparks from this, more than an inch long, if the weather was dry. (Nothing worked if the weather was damp, for then everything conducted.)

One could connect the electrical machine to a Leyden jar—

basically a glass jar coated with tinfoil on both sides, and a metal ball at the top, connected to the inside foil by a metal chain. If one connected several such jars together, they could hold a formidable charge. It was such a "battery" of Leyden jars in the eighteenth century, I read, which had been used in one experiment to give an almost paralyzing shock to a line of eight hundred soldiers, all of them joined by holding hands.

I also got a small Wimshurst machine, a beautiful thing with revolving glass discs and radiating metal sectors that could yield massive sparks up to four inches long. When the plates of the Wimshurst machine were revolving fast, everything around it became highly charged: tassels became electrified, their threads straining apart; pithballs would fly apart, and one felt the electricity on one's skin. If there was a sharp point nearby, electricity would stream from it in a luminous brush, a little corposant, and one could blow out candles with the outstreaming "electric wind," or even get this to turn a little rotor on its pivot. Using a simple insulating stool—a wooden board supported by four tumblers—I was able to electrify my brothers so their hair stood on end. These experiments showed the repulsive power of like electric charges, each thread of the tassel, each hair, acquiring the same charge (whereas my first experience, with rubbed amber and bits of paper, had shown the power of electrically charged bodies to attract). Opposites attracted, likes repelled.

I wondered whether one could use the static electricity of the Wimshurst machine to light up one of Uncle Dave's lightbulbs. Uncle said nothing, but provided me with some very fine wire made of silver and gold only a three-hundredth of an inch thick. When I connected the brass balls of the Wimshurst machine with a three-inch length of silver wire on a card, the wire exploded when I turned the handle, leaving a strange pattern on the card. And when I tried it with the gold wire, this was vaporized instantly, turning into a red vapor, gaseous gold. It seemed to me from these experiments that frictional electricity could be quite formidable—but that it was too violent, too intractable, to be of much use.

Electrochemical attraction, for Davy, was the attraction of opposites—the attraction, for example, of an intensely "positive" metallic ion, a cation like that of sodium, to an intensely "negative" one, an anion like that of chloride. But most elements, he thought, came between these on a continuous scale of electropositivity or -negativity. The degree of electropositivity among metals went with their chemical reactivity, hence their ability to reduce or replace less positive elements.

This sort of replacement, without any clear notion of its rationale, had been explored by the alchemists in the production of metallic coatings or "trees." Such trees were made by inserting a stick of zinc, say, into a solution of another metallic salt (a silver salt, for example). This would result in the displacement of the silver by the zinc, and metallic silver would be precipitated from the solution as a shining, almost fractal, arborescent growth. (The alchemists had given these trees mythical names, so the silver tree was called Arbor Dianae, the lead tree Arbor Saturni, and the tin tree Arbor Jovis.)[1]

I had hoped, at one point, to make such trees of all the metallic elements—trees of iron and cobalt, and bismuth and nickel, of gold, of platinum, of all the platinum metals; of chromium and molybdenum, and (of course!) tungsten; but various considerations (not least, the prohibitive cost of the precious metal salts) confined me to a dozen or so basic ones. But the sheer aesthetic delight of these—no two trees ever looked the same; they

---

[1] These names for metallic trees came from the alchemical notion of the correspondence between the sun, the moon, and the five (known) planets with the seven metals of antiquity. Thus gold stood for the sun, silver for the moon (and the moon goddess, Diana), mercury for Mercury, copper for Venus, iron for Mars, tin for Jupiter (Jove), and lead for Saturn.

were as different, even with the same metal, as snowflakes or ice crystals, and different metals, one could see, were deposited in different ways—soon gave way to a more systematic study. When did one metal lead to the deposition of another? And why? I used a zinc rod, sticking it first into a solution of copper sulfate, and got a gorgeous encrustation, a copper plating, all around it. I then experimented with tin salts, lead salts, and silver salts, putting a zinc rod into solutions of these, and produced shining, crystalline trees of tin, lead, and silver. But when I tried to make a zinc tree, by sticking a copper rod into a solution of zinc sulfate, nothing happened. Zinc was clearly the more active metal, and as such could replace the copper, but not be replaced by it. To make a zinc tree, one had to use a metal even more active than zinc—a magnesium rod, I found, worked well. Clearly all these metals did form a sort of series.

Davy himself pioneered the use of electrochemical displacement for protecting the copper bottoms of ships from corrosion in seawater, attaching to them plates of more electropositive metals (such as iron or zinc), so that these would become corroded instead, a so-called cathodic protection. (Though this seemed to work well under laboratory conditions, it did not work well at sea, because the new metal plates attracted barnacles—and thus Davy's suggestion was ridiculed. Yet the principle of cathodic protection was brilliant, and eventually became, after his death, a standard way of protecting the bottoms of ocean-going vessels.)

Reading about Davy and his experiments stimulated me to a variety of other electrochemical experiments: I put an iron nail in water, attaching a piece of zinc to it to protect it from corrosion. I removed the tarnish from my mother's silver spoons by putting them in an aluminum dish with a warm solution of sodium bicarbonate. She was so pleased by this that I decided to go further and try electroplating, using chromium as the anode and a variety of household objects as the cathode. I chromium-plated everything I could lay hands on—iron nails, bits of copper, scissors, and (this time to my mother's considerable annoyance) one of the silver spoons that I had previously cleaned of tarnish.

I did not realize at first that there was any connection between these experiments and the batteries I was playing with at the same time, although I thought it an odd coincidence that the first pair of metals I used, zinc and copper, could produce either a tree or, in a battery, an electric current. I think it was only when I read that, to get a higher voltage, batteries used nobler metals such as silver and platinum that I started to realize that the two series—the "tree" series and Volta's series—were probably the same, that chemical activity and electrical potential were in some sense the same phenomenon.

We had a large old-fashioned battery, a wet cell, in the kitchen, hooked up to an electric bell. The bell was too complicated to understand at first, and the battery, to my mind, was more immediately attractive, for it contained an earthenware tube with a massive, gleaming copper cylinder in the middle, immersed in a bluish liquid; all this inside an outer glass casing, also filled with fluid, and containing a slimmer bar of zinc. It looked like a miniature chemical factory of sorts, and I thought I saw little bubbles of gas, at times, coming off the zinc. This Daniell cell (as it was called) had a thoroughly nineteenth-century, Victorian look about it, and this extraordinary object was making electricity all by itself—not by rubbing or friction, but just by virtue of its own chemical reactions. That this was quite another source of electricity, not frictional or static, but a radically different *sort* of electricity, must have seemed astounding in the extreme, a new force of nature, when Volta discovered it in 1800. Previously there had been only the fugitive discharges, the sparks and flashes, of frictional electricity; now one could have at one's disposal a steady, uniform, unvarying current. One only needed two different metals—copper and zinc would do, or copper and silver (Volta worked out a whole series of metals, differing in the "voltage," the potential difference, between them), immersed in a conducting medium.

The first batteries I made myself used fruit or vegetables—one could stick copper and zinc electrodes into a potato or a lemon and get enough current to light a tiny 1-volt bulb. And one could wire half a dozen lemons or potatoes together (in series to get a higher voltage, or in parallel to get more power) to make a biological "battery." After the fruit and vegetable batteries, I turned to coins, using alternating copper and silver coins (one had to use silver coins made before 1920, for later ones were debased) with moistened (usually saliva-moistened) blotting paper between them. If I used small coins, farthings and sixpences, I could get five or six such couples in an inch, or I could make a pile a foot high, with sixty or seventy couples, enclosed in a tube, which could give quite a sharp, 100-volt shock. One could go on, I thought, to make an electric stick filled with narrow couples of copper and zinc foil, a lot thinner than coins. Such a stick, with five hundred or more couples, might generate a thousand volts, more even than an electric eel, enough to frighten off any assailant—but I never got as far as making one.

I was fascinated by the huge range of batteries developed in the nineteenth century, some of which I could see in the Science Museum. There were "single fluid" batteries, like Volta's original cell, or the Smee, or the Grenet, or the massive Leclanche, or the slim, silver battery of de la Rue; and there were two-fluid batteries, like our own Daniell, and the Bunsen, and the Grove (which used platinum electrodes). Their number seemed endless, but all were designed, in their different ways, to secure a more reliable and constant flow of current, to protect the electrodes from the deposition of metal or the adherence of gas bubbles, and to avoid (as some batteries caused) the emission of noxious or inflammable gases.

These wet cells had to be topped up with water from time to

time; but the little dry cells in our torches were clearly different. Marcus, seeing my interest, dissected one for me, using his powerful scout knife, showing me the outer case of zinc, the central carbon rod, and the rather corrosive and strange-smelling conducting paste between them. He showed me the massive 120-volt battery in our portable radio (this was a necessity in the war, when the electricity supply was so erratic)—it contained eighty linked dry cells, and weighed several pounds. And once he opened the bonnet of the car—we had the old Wolseley at the time—and showed me the accumulator, with its lead plates and acid, and explained how this had to be charged, and could carry a charge repeatedly, but not generate one itself. I adored batteries, and they did not have to be live; when my interest was made known to the family, used batteries of all shapes and sizes poured in, and I rapidly accumulated a remarkable (though wholly useless) collection of the things, many of which I opened and dissected.

But my favorite remained the old Daniell cell, and when we went modern and got a natty new dry cell for the bell, I appropriated the Daniell for myself. It had only a modest voltage of 1 or 1½ volts, but the current, several amperes, was considerable in view of its size. This made it very suitable for heating and lighting experiments, where one needed a substantial current, but the voltage hardly mattered.

Thus I could readily heat wire—Uncle Dave had supplied me with a whole bandolier of fine tungsten wire of all different thicknesses. The thickest wire, two millimeters in diameter, became mildly warm when I connected a length of it across the terminals of the cell; the thinnest wire grew white-hot and incinerated in a flash; there was a comfortable in-between wire that one could maintain for a little while at red heat, though even at this temperature it soon oxidized and disintegrated into a fluff of yellowish white oxide. (Now I knew why it had been crucial to remove the air from lightbulbs, and why incandescent lighting was not possible unless the bulbs were evacuated or filled with an inert gas.)

I could also, using the Daniell as a source of power, decompose water if it was briny or acidulated. I remember the extraordinary pleasure I got from decomposing a little water in an eggcup, seeing it visibly separate into its elements, oxygen at one electrode, hydrogen at the other. The electricity from a 1-volt cell seemed so mild, and yet it could suffice to tear a chemical compound apart, to decompose water or, more dramatically, salt into its violently active constituents.

Electrolysis could not have been discovered before Volta's pile, for the most powerful electrical machines or Leyden jars were wholly impotent to cause chemical decomposition. It would have required, Faraday later calculated, the massed charge of 800,000 Leyden jars, or perhaps the power of a whole lightning stroke, to decompose a single grain of water, something that could be done by a tiny and simple 1-volt cell. (But my 1-volt cell, on the other hand, or even the eighty-cell battery that Marcus showed me in the portable radio, could not make a pithball or an electroscope move.) Static electricity could generate great sparks and high-voltage charges (a Wimshurst machine could generate 100,000 volts), but very little power, at least to electrolyze. And the opposite was so with the massive power, but low voltage, of a chemical cell.

If the electric battery was my introduction to the inseparable relation of electricity to chemistry, the electric bell was my introduction to the inseparable relation of electricity to magnetism — a relation by no means self-evident or transparent, and one that was discovered only in the 1820s.

I had seen how a modest electric current could heat a wire, give a shock, or decompose a solution. How was it managing to cause the oscillating movement, the clatter, of our electric bell? Wires from the bell ran to the front door, and a circuit was completed when the outside button was pressed. One evening when my parents were out, I decided to bypass this circuit, and connected the wires so that I could actuate the bell directly. As soon as I let the current pass, the bell hammer jumped, hitting the bell. What made it jump when the current flowed? I saw how the bell hammer, which was made of iron, had copper wire coiled around it. The coil became magnetized when a current flowed through it, and this caused the hammer to be attracted to the iron base of the bell (once it hit the bell, it broke the circuit and fell back into its original place). This seemed extraordinary to me: my lodestones, my horseshoe magnets, were one thing, but here was magnetism that appeared only when a current flowed through the coil, and disappeared the moment it stopped.

It was the delicacy, the responsiveness, of compass needles which had first given a clue to the connection between electricity and magnetism. It was well known that a compass needle might jerk or even get demagnetized in a thunderstorm, and in 1820 it was observed that if a current was allowed to flow through a wire near a compass, its needle would suddenly move. If the current was strong enough, the needle could be deflected ninety degrees. If one put the compass above the wire rather than below it, the needle turned in the opposite direction. It was as if the magnetic force were forming circles around the wire.[2]

---

[2] A discovery that for some reason especially interested me was Faraday's discovery of diamagnetism in 1845. He had been experimenting with a very powerful new electromagnet, placing various transparent substances between its poles to see whether polarized light could be affected by the magnet. It could, and Faraday now found that the very heavy lead glass that he had used for some experiments actually moved when the magnet was switched on, aligning itself at right angles to the magnetic field (this was the first time he used the term *field*). Prior to this all known magnetic substances—iron, nickel, magnetite, etc.—had aligned themselves along the magnetic field, rather than at right angles to it. Intrigued, Faraday went on to test the magnetic susceptibility of everything he could lay his hands on—not only metals and minerals, but glass, flames, meat, and fruit, too.

Such a circular movement of magnetic forces could readily be made visible by using a vertical magnet sticking in a bowl of mercury, with a loosely suspended wire just touching the mercury, and a second bowl in which the magnet could move and the wire was fixed. When a current flowed, the loosely suspended wire would skitter in circles around the magnet, and the loose magnet would rotate in the opposite direction around the fixed wire.

Faraday, who in 1821 designed this apparatus—in effect, the world's first electric motor—immediately wondered about its reverse: if electricity could produce magnetism so easily, could a magnetic force produce electricity? Remarkably, it took him several years to answer this question, for the answer was not simple.[3] Putting a permanent magnet inside a coil of wire did not gener-

---

When I spoke of this to Uncle Abe, he allowed me to experiment with the very powerful electromagnet he had in his attic, and I was able to duplicate a lot of Faraday's findings, and to find, as he had, that the diamagnetic effect was especially powerful with bismuth, which was strongly repelled by both poles of the magnet. It was fascinating to see how a thin shard of bismuth (as near a needle as I could get with the brittle metal) aligned itself, almost violently, perpendicular to the magnetic field. I wondered whether, if it was sufficiently delicately poised, one might make a bismuth compass that pointed east-west. I experimented with bits of meat and fish, and wondered about experimenting with living creatures, too. Faraday himself had written, "If a man could be in the magnetic field, like Mahomet's coffin he would turn until across the magnetic field." I wondered about putting a small frog, or perhaps an insect, in the field of Uncle Abe's magnet, but feared this might freeze the motion of its blood, or blow its nervous system, turn out to be a refined form of murder. (I need not have worried: frogs have now been suspended for minutes in magnetic fields, and are apparently none the worse for the experience. With the vast magnets now available, an entire regiment could be suspended.)

[3] He was distracted, too, creatively, by a dozen competing interests and commitments during this time: the investigation of steels, the making of special highly refractive optical glasses, the liquefaction of gases (which he was the first to achieve), the discovery of benzene, his many chemical and other lectures at the Royal Institution, and the publication in 1827 of his *Chemical Manipulations*.

ate any electricity; one had to move the bar in and out, and only then was a current generated. It seems obvious to us now, because we are familiar with dynamos and how they work. But there was no reason at the time to expect that movement would be necessary; after all, a Leyden jar, a voltaic battery, just sat on the table. It took even a genius like Faraday ten years to make the mental leap, to move out of the assumptions of his time into a new realm, and to realize that movement of the magnet was necessary to generate electricity, that movement was of the essence. (Movement, Faraday thought, generated electricity by cutting the magnetic lines of force.) Faraday's in-and-out magnet was the world's first dynamo—an electric motor in reverse.

It was curious that Faraday's two inventions, the electric motor and the dynamo, discovered around the same time, had very different impacts. Electric motors were taken up and developed almost at once, so that there were battery-powered electric riverboats by 1839, while dynamos were much slower to develop and became widespread only in the 1880s, when the introduction of electric lights and electric trains created a demand for huge amounts of electricity and a distribution system to keep them going. Nothing like these vast, humming dynamos, weaving a mysterious and invisible new power out of thin air, had ever been seen, and the early powerhouses, with their great dynamos, inspired a sense of awe. (This is evoked in H. G. Wells's early story "The Lord of the Dynamos," in which a primitive man begins to see the massive dynamo he looks after as a god who demands a human sacrifice.)

Like Faraday, I started to see "lines of force" everywhere. I already had battery-powered front and rear lamps on my bike, and now I got dynamo-powered lights as well. As the little dynamo whirred on the back wheel, I would think sometimes of the magnetic lines of force being cut as it whirred, and of the mysterious, crucial role of motion.

Magnetism and electricity had seemed at first completely separate; now they seemed to be linked, somehow, by motion. It was at this point that I turned to my "physics" uncle, Uncle Abe, who explained that the relationship between electricity and magnetism (and the relationship of both to light) had indeed been made clear by the great Scottish physicist Clerk Maxwell.[4] A moving electrical field would induce a magnetic field near it, and this in turn would induce a second electrical field, and this another magnetic field, and so on. With these almost instantaneous mutual inductions, Maxwell envisaged, there would be, in effect, a combined electromagnetic field in extremely rapid oscillation, and this would expand in all directions, propagating itself as a wave motion through space. In 1865, Maxwell was able to calculate that such fields would propagate at 300,000 kilometers per second, a velocity extremely close to that of light. This was very startling—no one had suspected any relationship between magnetism and light; indeed, no one had any idea what light might be, although it was well understood that it was propagated as a wave. Now Maxwell suggested that light and magnetism were "affections of the same substance, and that light is an electromagnetic disturbance propagated through the field according to

---

[4] Having no higher mathematics myself, unlike Uncle Abe, I found much of Maxwell's work inaccessible, whereas I could at least read Faraday and feel I was getting the essential ideas, despite the fact that he never used mathematical formulas. Maxwell, expressing his indebtedness to Faraday, spoke of how his ideas, though fundamental, could be expressed in nonmathematical form:

It was perhaps for the advantage of science that Faraday, though thoroughly conscious of the fundamental forms of space, was not a professed mathematician . . . and did not feel called upon . . . to force his results into a shape acceptable to the mathematical taste of the time. . . . He was thus left at leisure to do his proper work, to coordinate his ideas with the facts, and to express them in natural untechnical language. . . . [Yet, Maxwell continued] As I proceeded with the study of Faraday I perceived that his method of conceiving the phenomena was also a mathematical one, though not exhibited in the conventional form of mathematical symbols.

electromagnetic laws." After hearing this, I began to think of light differently—as electric and magnetic fields leapfrogging over each other with lightning speed, braiding themselves together to form a ray of light.

It followed, as a corollary, that any varying electric or magnetic field could give rise to an electromagnetic wave propagating in all directions. It was this, Abe said, that inspired Heinrich Hertz to look for other electromagnetic waves—waves, perhaps, with a much longer wavelength than visible light. He was able to do this, in 1886, by using a simple induction coil as a "transmitter" and small coils of wire with tiny (a hundredth of a millimeter) spark gaps as "receivers." When the induction coil was set to sparking, he could observe, in the darkness of his lab, tiny secondary sparks in the small coils. "You switch on the wireless," said Abe, "and you never think of the wonder of what's actually happening. Think how it must have seemed on that day in 1886 when Hertz saw these sparks in the darkness and realized that Maxwell was right, and that something like light, an electromagnetic wave, was raying out from his induction coil in every direction."

Hertz died as a very young man, and never knew that his discovery was to revolutionize the world. Uncle Abe himself was only eighteen when Marconi first transmitted radio signals across the English Channel, and he remembered the excitement of this, even greater than the excitement over the discovery of X-rays two years earlier. Radio signals could be picked up by certain crystals, especially crystals of galena; one would have to find the right spot on their surface by exploring them with a tungsten wire, a "cat's whisker." One of Uncle Abe's own early inventions was to make a synthetic crystal that worked even better than galena. Everyone still spoke of radio waves as "Hertzian waves" at this point, and Abe had called his crystal Hertzite.

But the supreme achievement of Maxwell was to draw all electromagnetic theory together, to formalize it, to compress it, into just four equations. In this half-page of symbols, Abe said, show-ing the equations to me in one of his books, was condensed the whole of Maxwell's theory—for those who could understand them. Maxwell's equations revealed, for Hertz, the lineaments of "a new physics . . . like an enchanted fairyland"—not only the possibility of generating radio waves, but a sense that the whole universe was crisscrossed by electromagnetic fields of every sort, reaching to the ends of the universe.

15

HOME LIFE

Zionism played a considerable part on both sides of my family. My father's sister Alida worked during the Great War as an assistant to Nahum Sokolov and Chaim Weitzmann, the leaders of Zionism in England at the time, and, with her gift for languages, was entrusted with the translation of the Balfour Declaration in 1917 into French and Russian, and her son Aubrey, even as a boy, was a learned and eloquent Zionist (and later, as Abba Eban, the first Israeli ambassador to the United Nations). My parents, as doctors with a large house, were expected to provide a venue, a hospitable place, for Zionist meetings, and such meetings often took over the house in my childhood. I would hear them from my bedroom upstairs—raised voices, endless argument, passionate poundings of the table—and every so often a Zionist, flushed with anger or enthusiasm, would barge into my room, looking for the loo.

These meetings seemed to take a lot out of my parents—they would look pale and exhausted after each one—but they felt a duty to host them. I never heard them talk between themselves about Palestine or Zionism, and I suspected they had no strong convictions on the subject, at least until after the war, when the horror of the Holocaust made them feel there should be a "National Home." I felt they were bullied by the organizers of these meetings, and by the gangsterlike evangelists who would pound at the front door and demand large sums for yeshivas or "schools in Israel." My parents, clearheaded and independent in most other ways, seemed to become soft and helpless in the face of these demands, perhaps driven by a sense of obligation or anxiety. My own feelings (which I never discussed with them) were passionately negative: I came to hate Zionism and evangelism and politicking of every sort, which I regarded as noisy and intrusive and bullying. I longed for the quiet discourse, the rationality, of science.

My parents were moderately orthodox in practice (though there was little discussion that I remember as to what anyone actually believed), but some of the family were extremely orthodox. It was said that my mother's father would wake up at night if his yarmulke fell off, and that my father's father would not even swim without his. Some of my aunts wore *sheitls*—wigs—and these gave an oddly youthful, sometimes mannequinlike appearance to them: Ida had a bright yellow one, Gisela a raven black one, and these remained unchanged even when my own hair, many years later, had started to turn grey.

My mother's eldest sister, Annie, had gone to Palestine in the 1890s and founded a school in Jerusalem, a school for "English gentlewomen of the Mosaic persuasion." Annie was a woman of commanding presence. She was excessively orthodox, and (I suspect) believed herself to be on close personal terms with the Deity (as she was with the Chief Rabbi, the Mandate, and the Mufti, in Jerusalem).[1] She would arrive periodically in England with steamer trunks so enormous they needed six porters to lift them, and on her visits she would introduce an atmosphere of terrifying religious strictness in the house—my parents, less orthodox, were somewhat scared of her gimlet eye.

On one occasion—it was an oppressive Saturday in the tense summer of 1939—I decided to ride my tricycle up and down Exeter Road near the house, but there was a sudden downpour and I got completely soaked. Annie wagged a finger at me, and shook her heavy head: "Riding on shabbas! You can't get away with it," she said. "He sees everything, He is watching all the time!" I disliked Saturdays from this time on, disliked God, too (at least the vindictive, punitive God that Annie's warning had evoked), and developed an uncomfortable, anxious, watched feeling about Saturdays (which persists, a little, to this day).

---

[1] Sir Ronald Storrs, the British governor of Jerusalem at the time, described his first encounter with Annie in his 1937 memoir, *Orientations:*

> When, early in 1918, a lady, unlike the stage Woman of Destiny in that she was neither tall, dark nor thin, was ushered, with an expression of equal good humour and resolution, into my office I immediately realized that a new planet had swum into my ken. Miss Annie Landau had been throughout the War exiled . . . from her beloved . . . girls' school, and demanded to return to it immediately. To my miserable pleading that her school was in use as a military hospital she opposed a steely insistence: and very few minutes had elapsed before I had leased her the vast empty building known as the Abyssinian Palace. Miss Landau rapidly became very much more than the headmistress of the best Jewish girls' school in Palestine. She was more British than the English . . . she was more Jewish than the Zionists—no answer from her telephone on the Sabbath, even by the servants. She had been friendly with the Turks and Arabs before the War; so that her generous hospitality was for many years almost the only neutral ground upon which British officials, ardent Zionists, Moslem Beys and Christian Effendis could meet on terms of mutual conviviality.

In general—that Saturday was an exception—I would go with the family to shul, the commodious Walm Lane Synagogue which, at that time, had a congregation of over two thousand. We would all be scrubbed and excessively clean, and dressed in our "Sunday" best, and walk down Exeter Road following our parents, like so many ducklings. My mother, along with various aunts, would climb to the women's gallery. When I was very young, three or less, I would go with her, but as a "grown-up" boy of six, I was expected to be downstairs with the men (though I was always stealing glances at the women upstairs, and sometimes tried to wave, though I was sternly forbidden to do so).

My father was well known in the congregation—half of whom were his patients, or my mother's—and had the reputation of being a staunch supporter of the community and a scholar, though his scholarship was nothing, he told me, to that of Wilensky across the aisle, who knew every word of the Talmud so thoroughly by heart that if a pin were stuck into any of the volumes, he could tell you what sentence it would pierce on every page. Wilensky did not follow the service, but some internal program or litany of his own, always rocking back and forth, davening, in his own way. He had long ringlets, and *payes* down his face—I looked at him with awe, as something superhuman.

It was a very long service on Saturday mornings, which even with high-speed praying took a minimum of three hours—and the praying was, at times, incredibly fast. One silent prayer, the Amidah, had to be said standing, facing toward Jerusalem. It was, I supposed, about ten thousand words long, but the front-runners in the shul could do it in three minutes flat. I would read as much as I could (with frequent glances at the translation on the opposite page to see what it all meant), but I had scarcely read more than a paragraph or two before the time was up, and the service rushed ahead onto something else. For the most part I did not try to keep up, but wandered through the prayer book in my own way. It was here that I learned about myrrh and frankincense, and the weights and measures used in the land of Israel three thousand years ago. There were many passages which attracted me with their rich language, or their beauty, their sense of poetry and myth, detailing the odors and spices that went with some sacrifices. It was evident that God had an acute nose.[2]

I liked the singing, the choir—where cousin Dennis sang, and Uncle Moss presided—the virtuosic *chazzan,* and some of the savage, rabbinical speeches, and occasionally the sense that all of us actually formed a single community. But by and large, the synagogue oppressed me; religion seemed more real, and infinitely more pleasant, at home. I loved Passover, with its preliminaries (removing all the leavened bread, the *chometz,* from the house, burning it, sometimes communally with our neighbors), the special, beautiful cutlery and plates and tablecloths we used for its eight days, and the rooting up of the horseradish that had been growing in the garden, its grinding which led to copious tears.

We would sit down fifteen, sometimes twenty, to the table on seder nights: my parents; the maiden aunts—Birdie, Len, and before the war, Dora, sometimes Annie; cousins of varying degree, visiting from France or Switzerland; and always a

---

[2] "The compound forming the incense," the Talmud prescribed in almost stoichiometric terms,

> consisted of balm, onycha, galbanum and frankincense, in quantities weighing seventy manehs each; of myrrh, cassia, spikenard and saffron, each sixteen manehs by weight; of costus twelve, of aromatic bark three, and of cinnamon nine manehs; of lye obtained from a species of leek, nine kabs; of Cyprus wine three seahs and three kabs: though, if Cyprus wine was not procurable, old white wine might be used; of salt of Sodom the fourth part of a kab, and of the herb Maaleh Ashan a minute quantity. R. Nathan says, a minute quantity was also required of the odoriferous herb Cippath, that grew on the banks of the Jordan; if, however, one added honey to the mixture, he rendered the incense unfit for sacred use, while he who, in preparing it, omitted one of its necessary ingredients, was liable to the penalty of death.

stranger or two who would come. There was a beautiful, embroidered tablecloth which Annie had brought us from Jerusalem, gleaming white and gold on the table. My mother, knowing that sooner or later there would be accidents, always had a preemptive "spill" herself—she would manage, somehow, very early in the evening, to tip a bottle of red wine onto the tablecloth, and thereafter no guest would be embarrassed if they knocked over a glass. Though I knew she did this deliberately, I could never predict how or when the "accident" would occur; it always looked absolutely spontaneous and authentic. (She would immediately spread salt on the wine stain, and it became much paler, almost disappearing; I wondered why salt had this power.)

Unlike the shul service, which was gabbled as fast as possible, and largely unintelligible to me, the seder service took its time, with long discussions and disquisitions, and questions about the symbolism of the different dishes—the egg, the salt water, the bitter herb, the *haroseth*. The Four Boys mentioned in the service—the Wise One, the Wicked One, the Simple One, and the One Who Was Too Young to Ask Any Questions—were always identified by me with the four of us, though this was especially unfair to David, who was neither more nor less wicked than any other fifteen-year-old boy. I loved the ritual washing of the hands, the four cups of wine, the recitation of the ten plagues (here, as one recited them, one would dip an index finger into the wine at each plague; then, after the tenth plague, the slaying of the firstborn, one would throw the wine on one's fingertips over one's shoulder). I, as the youngest, would recite the Four Questions in a quavering treble; and later, try to see where my father hid the middle matzoh, the *afikomen* (but I could no more catch him doing this than I could catch my mother maneuvering the wine spill).

I loved the songs and recitations of the seder, the feeling of a remembering, a ritual, which had been performed for millennia—the story of the bondage in Egypt, the infant Moses in the bulrushes being rescued by the pharaoh's daughter, the Promised Land flowing with milk and honey. I would be transported, we all would, into a mythic realm.

The seder service would go on past midnight, sometimes to one or two in the morning, and as a five- or six-year-old, I would be nodding off. Then, when it finally broke up, another cup of wine—the fifth cup—would be left for "Elijah" (he would come in the night, I was told, and drink the wine left for him). Since my own Hebrew name was Eliahu, Elijah, I decided that I was entitled to drink the wine, and in one of the last seders before the war, I slipped down at night and drank the whole cup. I was never questioned, and never admitted what I had done, but my hangover the next morning, and the empty cup, made any confession unnecessary.

I enjoyed all the Jewish festivals in different ways, but Succoth, the harvest festival, especially, for here we would build a house of leaves and branches, a *succah,* in the garden, its roof hung with vegetables and fruit, and if weather permitted, I could sleep in the *succah,* and look through the fruit-hung roof at the constellations above me.

But the more serious festivals, and the fasts, took me back to the oppressive atmosphere of the synagogue, an atmosphere that reached a sort of horror on the Day of Atonement, Yom Kippur, when all of us (we understood) were being weighed in the balance. One had ten days between New Year and the Day of Atonement to repent and make restitution for one's misdemeanors and sins, and this repentance reached its climax, communally, on Yom Kippur. During this time, of course, we had all been fasting, no food or drink being allowed to pass our lips for twenty-five hours. We would beat our breasts and wail: "We have done this, we have done that"—all possible sins were mentioned (including many I had never thought of), sins of commission and omission, sins deliberate and inadvertent. The terrifying thing was that one did not know whether one's breast-beating was convincing to God, or whether one's sins were even, in fact, forgivable. One did not know whether He would reinscribe one in the Book of Life, as the liturgy had it, or whether one would die and be cast into outer darkness. The intense, tumultuous emotions of the congregation were expressed by the astonishing voice of our old *chazzan,* Schechter—Schechter, as a young man, had wanted to sing in opera, but never in fact sang outside the synagogue. At the very end of the service, Schechter would blow the shofar, and with this the atonement was over.

When I was fourteen or fifteen—I am not sure of the year— the Yom Kippur service ended in an unforgettable way, for Schechter, who always put great effort into the blowing of the shofar—he would go red in the face with exertion—produced a long, seemingly endless note of unearthly beauty, and then dropped dead before us on the *bema,* the raised platform where he would sing. I had the feeling that God had killed Schechter, sent a thunderbolt, stricken him. The shock of this for everyone was tempered by the reflection that if there was ever a moment in which a soul was pure, forgiven, relieved of all sin, it was at this moment, when the shofar was blown in conclusion of the fast; and that Schechter's soul, almost certainly, had fled its body at this moment and gone straight to God. It was a holy dying, everyone said: please God, when their time came, they might die like this too.

Strangely, both my grandfathers had, in fact, died on Yom Kippur (though not in circumstances quite as dramatic as these), and at the start of every Yom Kippur, my parents would light squat mourning candles for them, which would burn slowly throughout the fast.

In 1939 an older sister of my mother's, Auntie Violet, had come from Hamburg with her family. Her husband, Moritz, was a chemistry teacher and much-decorated veteran of the First World War, who had been wounded by shell fragments and walked with a heavy limp. He thought of himself as a patriotic German and could not believe that he would ever be forced to flee his native country, but Kristallnacht had finally brought home to him the fate that awaited him and his family if they did not escape, and in the spring of 1939 they made it to England—just (all their property had been seized by the Nazis). They stayed with Uncle Dave, and briefly with us, before going to Manchester, where they opened a school and hostel for evacuees.

Occupied, preoccupied, with my own state, I was largely ignorant of much that was going on in the world at large. I knew little, for example, about the evacuation of Dunkirk in 1940, after the fall of France, the frantic crowding of boats with the last refugees to escape the Continent. But in December of 1940, home from Braefield for the holidays, I found that a Flemish couple, the Huberfelds, were now living in one of the spare rooms at 37. They had escaped in a small boat, hours before the German forces had arrived, and had then almost been lost at sea. They did not know what had befallen their own parents, and it was from them that I first gained some idea of the chaos and the horror in Europe.

During the war the congregation was largely broken up—as the young men volunteered or were called up for the military, and hundreds of the children, like Michael and myself, were evacuated from London—and it was never really reconstituted after the war. A number of the congregants were killed, either fighting in Europe or through the bombing in London; others moved away from what had been, before the war, an almost exclusively Jewish, middle-class suburb. Before the war my parents (I, too) had known almost every shop and shopkeeper in Cricklewood: Mr. Silver in his chemist's shop, the grocer Mr. Bramson, the greengrocer Mr. Ginsberg, the baker Mr. Grodzinski, the kosher butcher Mr. Waterman—and I would see them all in their places in shul. But all this was shattered with the impact of the war, and then with the rapid postwar social changes in our corner of London. I myself, traumatized at Braefield, had lost touch with, lost interest in, the religion of my childhood. I regret that I was to lose it as early and as abruptly as I did, and this feeling of sadness or nostalgia was strangely admixed with a raging atheism, a sort of fury with God for not existing, not taking care, not preventing the war, but allowing it, and all its horrors, to occur.

Her Hebrew name was Zipporah ("bird"), but to us, to the family, she was always Auntie Birdie. It was never quite clear to me (or perhaps anyone) what had happened to Birdie in early life. There was talk of a head injury in infancy, but also of a congenital disorder, a defective thyroid gland, and she had to take large doses of thyroid extract throughout her life. Birdie had somewhat creased and folded skin, even as a young woman; she was of small stature and modest intelligence, the only one so handicapped among the otherwise gifted and robust children of my grandfather. But I am not sure that I regarded her as "handicapped"; to me she was just Auntie Birdie, who lived with us, was an essential part of the house, always there. She had her own room, next to my parents' room, filled with photos, postcards, tubes of colored sand, and knickknacks from family holidays going back to the beginning of the century. Her room had a clean, almost puppylike smell and was an oasis of calm for me, sometimes, when the house was in an uproar. She had a fat yellow Parker pen (my mother had an orange one), and wrote slowly in an unformed, childlike hand. I knew, of course, that there was "something wrong" with Birdie, something medically the matter, that her health was fragile and her powers of mind limited, but none of this really mattered, or was relevant to us. We knew only that she was there, a constant presence, unwaveringly devoted, and that she seemed to love us without ambivalence or reservation.

When I became interested in chemistry and mineralogy, she

would go out and get small mineral specimens for me; I never knew where or how she got these (nor how, after asking Michael what book I might like for my bar mitzvah, she got me a copy of Froissart's *Chronicles*). As a young woman, Birdie had been employed by the firm of Raphael Tuck, which published calendars and postcards, as one of an army of young women who painted and colored the cards — these delicately colored cards were very popular, and often collected, for decades, and seemed a permanent part of life until the 1930s, when color photography and color printing started to displace them, and to render Tuck's small army of women superfluous. In 1936, after almost thirty years of working for them, Birdie was dismissed one day, with no warning and scarcely a "thank you," let alone a pension or severance pay. When she came back that evening (Michael told me years later) her face was "stricken," and she never quite got over this.

Birdie was at once so quiet, so unassuming, so ubiquitous, that we all tended to take her for granted and to overlook the crucial role she played in our lives. When, in 1951, I got a scholarship to Oxford, it was Birdie who gave me the telegram, and hugged and congratulated me — shedding some tears, too, because she knew this meant that I would be leaving home.

Birdie had frequent attacks of "cardiac asthma," or acute heart failure, in the night, when she would get short of breath, and very anxious, and need to sit up. This sufficed at first for her milder attacks, but as they grew more severe, my parents asked her to keep a little brass bell by her bedside and to ring it as soon as she felt any distress. I would hear the little bell ring at increasingly frequent intervals, and it started to dawn on me that this was a serious condition. My parents would get up at once to treat Birdie — she needed oxygen now, and morphine, to get her through her attacks — and I would lie in bed, listening fearfully until all was calm again and I could return to sleep. One night, in 1951, the little bell rang, and my parents rushed into the room.

Her attack, this time, was extremely severe: pink froth was coming out of her mouth—she was drowning in the fluid that had welled into her lungs—and she did not respond to the oxygen and morphine. As a final, desperate measure to save her life, my mother performed a venesection with a scalpel on Birdie's arm, in an attempt to relieve the pressure on the heart. But it did not work with Birdie, and she died in my mother's arms. When I entered the room, I saw blood everywhere—blood all over her nightdress and arms, blood all over my mother, who was holding her. I thought for a moment that my mother had killed her, before I deciphered the fearful scene before me.

It was the first death of a close relative, of someone who had been an essential part of my life, and it affected me much more deeply than I had expected.

As a child, it seemed to me that the house was full of music. There were two Bechsteins, an upright and a grand, and sometimes both were being played simultaneously, to say nothing of David's flute and Marcus's clarinet. At such times the house was a veritable aquarium of sound, and I would become aware of one instrument and then another as I walked about (the different instruments did not seem to clash, curiously; my ear, my attention, would always select one or another).

My mother was not as musical as the rest of us, but very fond, nonetheless, of Brahms and Schubert lieder; she would sing these, sometimes, with my father accompanying her at the piano. She was especially fond of Schubert's "Nachtgesang," his Song of the Night, which she would sing in a soft, slightly off-key voice. This is one of my earliest memories (I never knew what the words meant, but the song affected me strangely). I cannot hear this now without recalling with almost unbearable vividness our drawing room as it was before the war, and my mother's figure and voice as she leaned over the piano and sang.

My father was very musical, and would come back from con-

certs and play much of the program by ear, transposing fragments into different keys, playing with them in different ways.
He had an omnivorous love of music, and enjoyed music halls as
much as chamber concerts, Gilbert and Sullivan as much as Mon-
teverdi. He was particularly fond of songs from the Great War,
and would sing these in a resounding bass. He had a large library
of miniature scores, and always seemed to have one or two of
these in his pockets (and indeed he usually went to bed with one
of them, or the dictionary of musical themes that I gave him later
for one of his birthdays).

Though he had studied with a noted pianist, and was always
darting to the keyboard of one or the other of the pianos, my
father's fingers were so broad and stubby that they could never fit
quite comfortably on the keys, so he usually contented himself
with impressionistic fragments. But he was eager for the rest of
us to be at home on the piano, and engaged a brilliant piano
teacher, Francesco Ticciati, for us all. Ticciati drilled Marcus and
David in Bach and Scarlatti with passionate, demanding intensity (Michael and I, younger, would play Diabelli duets), and at
times I would hear him bang the piano with frustration, shouting, "No! No! No!" when they failed to get things right. Then he
would sit down sometimes and play himself, and suddenly I
knew what *mastery* meant. He instilled in us an intense feeling for
Bach especially, and all the hidden structure of a fugue. When I
was five, I am told, and asked what my favorite things in the
world were, I answered, "smoked salmon and Bach." (Now, sixty
years later, my answer would be the same.)

I found the house somewhat stark, musicless, when I returned
to London in 1943. Marcus and David, pre-med students now,
were themselves evacuated—Marcus to Leeds, David to Lancaster; my father was busy, when not seeing patients, with his
duties as an air-raid warden; my mother equally so, doing emergency surgery late into the night at a hospital in St. Albans. I
would wait up, sometimes, to hear the sound of her bicycle bell,

as she cycled back, close to midnight, from the Cricklewood station.

A great treat at this time was to hear Myra Hess, the famous pianist who almost single-handed, it seemed, reminded Londoners in the midst of war of the timeless, transcendent beauties of music. We would often gather around the wireless in the lounge to hear the broadcasts of her lunchtime recitals.

When Marcus and David came back, after the war, to continue as medical students in London, the flute and the clarinet had long been abandoned, but David, it was evident, had exceptional musical gifts, was the one who really took after our father. David discovered blues and jazz, fell in love with Gershwin, and brought a new sort of music to our previously "classical" house. David was already a very good improviser and pianist, with a special flair for playing Liszt, but now suddenly the house was full of new names, names unlike any I had ever heard before: "Duke" Ellington, "Count" Basie, "Jelly Roll" Morton, "Fats" Waller— and from the horn of the new wind-up Decca gramophone he kept in his room I first heard the voices of Ella Fitzgerald and Billie Holiday. Sometimes when David sat down at the piano, I was not sure whether he was playing one of the jazz pianists or improvising something of his own—I think he wondered, half seriously, whether he might become a composer himself.

Both David and Marcus, I came to realize, though they seemed happy enough, and looked forward to being doctors, had a certain sadness, a sense of loss and renunciation, about other interests they had given up. For David this was music, while Marcus's passion, from an early age, had been for languages. He had an extraordinary aptitude for learning them, and was fascinated by their structure; at sixteen he was already fluent not only in Latin, Greek, and Hebrew, but in Arabic, which he had taught himself. He might have gone on, like his cousin Aubrey, to do oriental languages at university, but then the war came. Both he and David would have reached call-up age in 1941–42, and both became medical students, in part, to defer their call-up. But with this, I think, they deferred their other aspirations, a deferment that seemed permanent and irreversible by the time they returned to London.

Mr. Ticciati, our piano teacher, died in the war, and when I came back to London in 1943 my parents found another teacher for me, Mrs. Silver, a red-haired woman with a ten-year-old son, Kenneth, who was born deaf. After I had studied with her for a couple of years, she became pregnant once again. I had seen my mother's pregnant patients almost daily, as they came to her consulting room in the house, but this was the first time I saw someone so close to me go through an entire pregnancy. There were some problems toward the end—I heard talk of "toxemia" and I believe my mother had to do a "version" of the baby, so that it would come out head first. Finally Mrs. Silver went into labor and was admitted to hospital (my mother usually delivered babies at home, but here it seemed there might be complications and a caesarian section might be necessary). It did not occur to me that anything serious might happen, but when I got home from school that day, Michael told me that Mrs. Silver had died in childbirth, "on the table."

I was shocked, and outraged. How could a healthy woman die like this? How could my mother have let such a catastrophe occur? I never learned any details of what had happened, but the very fact that my mother had been present throughout evoked the fantasy that she had killed Mrs. Silver—even though everything I knew convinced me of my mother's expertise and concern, and that she must have encountered something beyond her power, beyond human power, to control.

I feared for Kenneth, Mrs. Silver's deaf son, whose main communication had been in a homemade sign language he shared only with his mother. And I lost the impulse to play the piano—I did not touch one at all for a year—and never allowed another piano teacher thereafter.

· · ·

I never thought I really knew or understood my brother Michael, even though he was the closest to me in age, and the one who came to Braefield with me. There is, of course, a great difference between six and eleven (our respective ages when we went to Braefield), but there seemed, in addition, something special about him which I (and perhaps others) were conscious of, though we would have found it difficult to characterize, much less to understand. He was dreamy, abstracted, deeply introspective; he seemed (more than any of us) to live in a world of his own, though he read deeply and constantly, and had the most amazing memory for his reading. He developed, when we were at Braefield, a particular preference for *Nicholas Nickleby* and *David Copperfield,* and knew the entire, immense books by heart, though he never explicitly compared Braefield to Dotheboys, or Mr. B. to the monstrous Dr. Creakle. But the comparisons were surely there, implicit, perhaps even unconscious, in his mind.

In 1941 Michael, now thirteen, left Braefield and went on to Clifton College, where he was unmercifully bullied. He made no complaints here, any more than he had ever complained about Braefield, but signs of trauma were visible to seeing eyes. Once, in the summer of 1943, soon after I had returned to London, Auntie Len, who was staying with us, spied Michael as he came, half naked, from the bath. "Look at his back!" she said to my parents, "it's full of bruises and wheals! If this is happening to his body," she continued, "what is happening to his mind?" My parents seemed surprised, said they had noticed nothing amiss, that they thought Michael was enjoying school, had no problems, was "fine."

Soon after this, Michael became psychotic. He felt a magical and malignant world was closing about him (I remember his telling me that the lettering had been "transformed" on the number 60 bus to Aldwych, so that the word *Aldwych* now appeared to be written in "old-witchy" letters like runes). He came to believe, very particularly, that he was "the darling of a flagellomaniac God," as he put it, subject to the special attentions of "a sadistic Providence." There was, again, no explicit reference to our flagellomaniac headmaster in Braefield, but I could not help feeling that Mr. B. was there, amplified, cosmified now to a monstrous Providence or God. Messianic fantasies or delusions appeared at the same time—if he was being tortured or chastised, this was because he was (or might be) the Messiah, the one for whom we had waited so long. Torn between bliss and torment, fantasy and reality, feeling he was going mad (or perhaps so already), Michael could no longer sleep or rest, but agitatedly strode to and fro in the house, stamping his feet, glaring, hallucinating, shouting.

I became terrified of him, for him, of the nightmare which was becoming reality for him, the more so as I could recognize similar thoughts and feelings in myself, even though they were hidden, locked up in my own depths. What would happen to Michael, and would something similar happen to me, too? It was at this time that I set up my own lab in the house, and closed the doors, closed my ears, against Michael's madness. It was at this time that I sought for (and sometimes achieved) an intense concentration, a complete absorption in the worlds of mineralogy and chemistry and physics, in science—focusing on them, holding myself together in the chaos. It was not that I was indifferent to Michael; I felt a passionate sympathy for him, I half-knew what he was going through, but I had to keep a distance also, create my own world from the neutrality and beauty of nature, so that I would not be swept into the chaos, the madness, the seduction, of his.

16

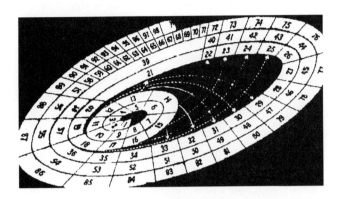

MENDELEEV'S GARDEN

In 1945 the Science Museum in South Kensington reopened (it had been closed for much of the war), and I first saw the giant periodic table displayed there. The table itself, covering a whole wall at the head of the stairs, was a cabinet made of dark wood with ninety-odd cubicles, each inscribed with the name, the atomic weight, and the chemical symbol of its element. And in each cubicle was a sample of the element itself (all of those elements, at least, which had been obtained in pure form, and which could be exhibited safely). It was labeled "The Periodic Classification of the Elements—after Mendeleeff."

My first vision was of metals, dozens of them in every possible form: rods, lumps, cubes, wire, foil, discs, crystals. Most were grey or silver, some had hints of blue or rose. A few had burnished surfaces that shone a faint yellow, and then there were the rich colors of copper and gold.

In the upper right corner were the nonmetals—sulfur in spectacular yellow crystals and translucent red crystals of selenium; phosphorus, like pale beeswax, kept under water; and carbon, as tiny diamonds and shiny black graphite. There was boron, a brownish powder, and ridged crystalline silicon, with a rich black sheen like graphite or galena.

On the left were the alkali and alkaline earth metals—the Humphry Davy metals—all (except magnesium) in protective baths of naphtha. I was struck by the lithium in the upper corner, for this, with its levity, was floating on the naphtha, and also by the cesium, lower down, which formed a glittering puddle beneath the naphtha. Cesium, I knew, had a very low melting point and it was a hot summer day. But I had not fully realized, from the tiny, partly oxidized lumps I had seen, that pure cesium was pale gold—it gave at first just a glint, a flash of gold, seeming to iridesce with a golden luster; then, from a lower angle, it was purely gold, and looked like a gilded sea, or golden mercury.

There were other elements which up to this point had only been names to me (or, almost equally abstract, names attached to some physical properties and atomic weights), and now for the first time I saw the range of their diversity and actuality. In this first, sensuous glance I saw the table as a gorgeous banquet, a huge table set with eighty-odd different dishes.

I had, by this time, become familiar with the properties of many elements and I knew they formed a number of natural families, such as the alkali metals, the alkaline earth metals, and the halogens. These families (Mendeleev called them "groups") formed the verticals of the table, the alkali and alkaline earth metals to the left, the halogens and inert gases to the right, and everything else in four intermediate groups in between. The "groupishness" of these intermediate groups was somewhat less clear—thus in Group VI, I saw sulfur, selenium, and tellurium. I knew that these three (my "stinkogens") were very similar, but what was oxygen doing, heading the group? There must be some deeper principle at work—and indeed there was. This was printed at the top of the table, but in my impatience to look at the elements themselves, I had paid no attention to it at all. The deeper principle, I saw, was valency. The term *valency* was not to be found in my early Victorian books, for it had only been properly developed in the late 1850s, and Mendeleev was one of the first to seize on it and use it as a basis for classification, to provide what had never been clear before: a rationale, a basis for the fact that elements seemed to form natural families, to have deep chemical and physical analogies with one another. Mendeleev now recognized eight such groups of elements in terms of their valencies.

Thus the elements in Group I, the alkali metals, had a valency of 1: one atom of these would combine with one atom of hydrogen, to form compounds such as LiH, NaH, KH, and so on. (Or with one atom of chlorine, to form compounds such as $LiCl$, $NaCl$, or KCl). The elements of Group II, the alkaline earth metals, had a valency of 2, and so would form compounds such as $CaCl_2$, $SrCl_2$, $BaCl_2$, and so on. The elements of Group VIII had a maximum combining power of 8.

But while Mendeleev was organizing the elements in terms of valency, he was also fascinated by atomic weights and the fact that these were unique and specific to each element, that they were, in a sense, the atomic signature of each element. And if, mentally, he started to index the elements according to their valencies, he did this equally in terms of their atomic weights. And now, magically, the two came together. For if he arranged the elements, quite simply, in order of their atomic weights, in horizontal "periods," as he called them, one could see recurrences of the same properties and valencies at regular intervals.

Every element echoed the properties of the one above, was a slightly heavier member of the same family. The same melody, so to speak, was played in each period—first an alkali metal, then an alkaline earth metal, then six more elements, each with its own valency or tone—but played in a different register (it was impossible to avoid thinking of octaves and scales here, for I lived in a musical house, and scales were the periodicity I heard daily).

It was eightness which dominated the periodic table before me, though one could also see, in the lower part of the table, that extra elements were interposed within the basic octets: ten extra elements apiece in Periods 4 and 5, and ten plus fourteen in Period 6.

So one went up, each period completing itself and leading to the next one in a series of dizzying loops—at least this is the form my imagination took, so that the sober, rectangular table before me was transformed, mentally, into spirals or loops. The table was a sort of cosmic staircase or a Jacob's ladder, going up to, coming down from, a Pythagorean heaven.

I got a sudden, overwhelming sense of how startling the periodic table must have seemed to those who first saw it—chemists profoundly familiar with seven or eight chemical families, but who had never realized the basis of these families (valency), nor how all of them might be brought together into a single overarching scheme. I wondered if they had reacted as I did to this first revelation: "Of course! How obvious! Why didn't I think of it myself?"

Whether one thought in terms of the verticals or in terms of the horizontals—either way one arrived at the same grid. It was like a crossword puzzle that could be approached by either the "down" or the "across" clues, except that a crossword was arbitrary, a purely human construct, while the periodic table reflected a deep order in nature, for it showed all the elements arrayed in a fundamental relationship. I had the sense that it harbored a marvelous secret, but it was a cryptogram without a key—*why* was this relationship so?

I could scarcely sleep for excitement the night after seeing the periodic table—it seemed to me an incredible achievement to have brought the whole, vast, and seemingly chaotic universe of chemistry to an all-embracing order. The first great intellectual clarifications had occurred with Lavoisier's defining of elements, with Proust's finding that elements combined in discrete proportions only, and with Dalton's notion that elements had atoms with unique atomic weights. With these, chemistry had come of age, and had become the chemistry of the elements. But the elements themselves were not seen to come in any order; they could only be listed alphabetically (as Pepper did in his *Playbook of Metals*) or in terms of isolated local families or groups. Nothing beyond this was possible until Mendeleev's achievement. To have perceived an *overall* organization, a superarching principle uniting and relating *all* the elements, had a quality of the miraculous, of genius. And this gave me, for the first time, a sense of the transcendent power of the human mind, and the fact that it might be equipped to discover or decipher the deepest secrets of nature, to read the mind of God.

I kept dreaming of the periodic table in the excited half-sleep of that night—I dreamed of it as a flashing, revolving pinwheel or Catherine wheel, and then as a great nebula, going from the first element to the last, and whirling beyond uranium, out to infinity. The next day I could hardly wait for the museum to open, and dashed up to the top floor, where the table was, as soon as the doors were opened.

On this second visit I found myself looking at the table in almost geographic terms, as a realm, a kingdom, with different territories and boundaries. Seeing the table as a geographic realm allowed me to rise above the individual elements, and see certain general gradients and trends. Metals had long been recognized as a special category of elements, and now one could see, in a single synoptic glance, how they occupied three-quarters of the realm—all of the west side, most of the south—leaving only a smallish area, mostly in the northeast, for the nonmetals. A

Periodic Table of the Elements

Key

74	Atomic number
W	Element's symbol
Tungsten	Element's name
183.85	Atomic weight

1 – Modern Group
IA – Mendeleevian Group

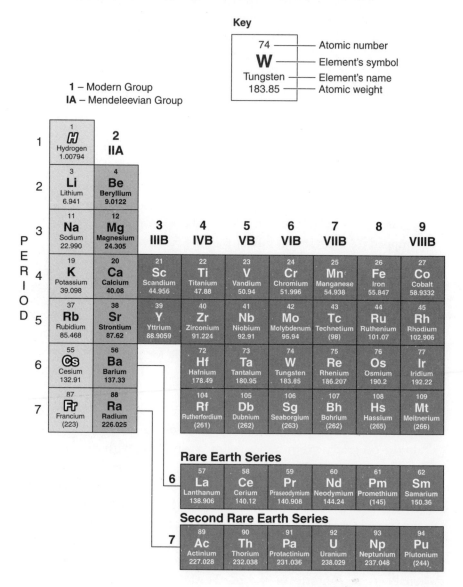

	1	2	3	4	5	6	7	8	9

P E R I O D

Period 1:
1 — **H** Hydrogen 1.00794
2 — **IIA**

Period 2:
3 — **Li** Lithium 6.941
4 — **Be** Beryllium 9.0122

Period 3:
11 — **Na** Sodium 22.990
12 — **Mg** Magnesium 24.305
3 — **IIIB**
4 — **IVB**
5 — **VB**
6 — **VIB**
7 — **VIIB**
8
9 — **VIIIB**

Period 4:
19 **K** Potassium 39.098
20 **Ca** Calcium 40.08
21 **Sc** Scandium 44.956
22 **Ti** Titanium 47.88
23 **V** Vanadium 50.94
24 **Cr** Chromium 51.996
25 **Mn** Manganese 54.938
26 **Fe** Iron 55.847
27 **Co** Cobalt 58.9332

Period 5:
37 **Rb** Rubidium 85.468
38 **Sr** Strontium 87.62
39 **Y** Yttrium 88.9059
40 **Zr** Zirconium 91.224
41 **Nb** Niobium 92.91
42 **Mo** Molybdenum 95.94
43 **Tc** Technetium (98)
44 **Ru** Ruthenium 101.07
45 **Rh** Rhodium 102.906

Period 6:
55 **Cs** Cesium 132.91
56 **Ba** Barium 137.33
72 **Hf** Hafnium 178.49
73 **Ta** Tantalum 180.95
74 **W** Tungsten 183.85
75 **Re** Rhenium 186.207
76 **Os** Osmium 190.2
77 **Ir** Iridium 192.22

Period 7:
87 **Fr** Francium (223)
88 **Ra** Radium 226.025
104 **Rf** Rutherfordium (261)
105 **Db** Dubnium (262)
106 **Sg** Seaborgium (263)
107 **Bh** Bohrium (262)
108 **Hs** Hassium (265)
109 **Mt** Meitnerium (266)

Rare Earth Series

6:
57 **La** Lanthanum 138.906
58 **Ce** Cerium 140.12
59 **Pr** Praseodymium 140.908
60 **Nd** Neodymium 144.24
61 **Pm** Promethium (145)
62 **Sm** Samarium 150.36

Second Rare Earth Series

7:
89 **Ac** Actinium 227.028
90 **Th** Thorium 232.038
91 **Pa** Protactinium 231.036
92 **U** Uranium 238.029
93 **Np** Neptunium 237.048
94 **Pu** Plutonium (244)

Alkali Metals

Alkaline Earth Metals

Transition Metals

Other Metals

Nonmetals

Halogens

Inert Gases

C Solid

Br Liquid

H Gas

Halogens

			13 IIIA	14 IVA	15 VA	16 VIA	17 VIIA	18 VIIIA
								2 He Helium 4.003
			5 B Boron 10.81	6 C Carbon 12.011	7 N Nitrogen 14.007	8 O Oxygen 15.999	9 F Fluorine 18.998	10 Ne Neon 20.179
10	11 IB	12 IIB	13 Al Aluminum 26.98	14 Si Silicon 28.086	15 P Phosphorus 30.974	16 S Sulfur 32.06	17 Cl Chlorine 35.453	18 Ar Argon 39.948
28 Ni Nickel 58.69	29 Cu Copper 63.546	30 Zn Zinc 65.39	31 Ga Gallium 69.72	32 Ge Germanium 72.59	33 As Arsenic 74.922	34 Se Selenium 78.96	35 Br Bromine 79.904	36 Kr Krypton 83.80
46 Pd Palladium 106.42	47 Ag Silver 107.868	48 Cd Cadmium 112.41	49 In Indium 114.82	50 Sn Tin 118.71	51 Sb Antimony 121.75	52 Te Tellurium 127.60	53 I Iodine 126.905	54 Xe Xenon 131.29
78 Pt Platinum 195.08	79 Au Gold 196.967	80 Hg Mercury 200.59	81 Tl Thallium 204.383	82 Pb Lead 207.2	83 Bi Bismuth 208.98	84 Po Polonium (209)	85 At Astatine (210)	86 Rn Radon (222)
110 (269)	111 (272)	112 (277)	113	114 (285)	115	116 (289)	117	118 (293)

63 Eu Europium 151.96	64 Gd Gadolinium 157.25	65 Tb Terbium 158.925	66 Dy Dysprosium 162.50	67 Ho Holmium 164.93	68 Er Erbium 167.26	69 Tm Thulium 168.934	70 Yb Ytterbium 173.04	71 Lu Lutetium 174.967
95 Am Americium (243)	96 Cm Curium (247)	97 Bk Berkelium (247)	98 Cf Californium (251)	99 Es Einsteinium (252)	100 Fm Fermium (257)	101 Md Mendelevium (258)	102 No Nobelium (259)	103 Lr Lawrencium (260)

jagged line, like Hadrian's Wall, separated the metals from the rest, with a few "semimetals," metalloids—arsenic, selenium—straddling the wall. One could see the gradients of acid and base, how the oxides of the "western" elements reacted with water to form alkalis, the oxides of the "eastern" elements, mostly non-metals, to form acids. One could see, again at a glance, how the elements on either border of the realm—the alkali metals and halogens, like sodium and chlorine, for example—showed the greatest avidity for each other and combined with explosive force, forming crystalline salts with high melting points which dissolved to form electrolytes; while those in the middle formed a very different sort of compound—volatile liquids or gases which resisted electric currents. One could see, remembering how Volta and Davy and Berzelius ranked the elements into an electrical series, how the most strongly electropositive elements were all to the left, the most strongly electronegative to the right. Thus it was not just the placement of the individual elements, but trends of every sort that hit the eye when one looked at the table.

Seeing the table, "getting" it, altered my life. I took to visiting it as often as I could. I copied it into my exercise book and carried it everywhere; I got to know it so well—visually and conceptually—that I could mentally trace its paths in every direction, going up a group, then turning right on a period, stopping, going down one, yet always knowing where I was. It was like a garden, the garden of numbers I had loved as a child—but unlike this, it was real, a key to the universe. I spent hours now, enchanted, totally absorbed, wandering, making discoveries, in the enchanted garden of Mendeleev.[1]

---

[1] Years later, when I read C. P. Snow, I found that his reaction to first seeing the periodic table was very similar to mine:

> For the first time I saw a medley of haphazard facts fall into line and order. All the jumbles and recipes and hotchpotch of the inorganic

There was a photograph of Mendeleev next to the periodic table in the museum; he looked like a cross between Fagin and Svengali, with a huge mass of hair and beard and piercing, hypnotic eyes. A wild, extravagant, barbaric figure—but as romantic, in his way, as the Byronic Humphry Davy. I needed to know more of him, and to read his famous *Principles,* in which he had first published his periodic table.

His book, his life, did not disappoint me. He was a man of encyclopedic interests. He was also a music lover and a close friend of Borodin (who was also a chemist). And he was the author of the most delightful and vivid chemistry text ever published, *The Principles of Chemistry.*[2]

Like my own parents, Mendeleev had come from a huge family—he was the youngest, I read, of fourteen children. His mother must have recognized his precocious intelligence, and when he reached fourteen, feeling that he would be lost without a proper education, she walked thousands of miles from Siberia with him—first to the University of Moscow (from which, as a Siberian, he was barred) and then to St. Petersburg, where he got a grant to train as a teacher. (She herself, apparently, nearing sixty at the time, died from exhaustion after this prodigious effort.

chemistry of my boyhood seemed to fit themselves into the scheme before my eyes—as though one were standing beside a jungle and it suddenly transformed itself into a Dutch garden.

[2] In his very first footnote, in the preface, Mendeleev spoke of "how contented, free, and joyous is life in the realm of science"—and one could see, in every sentence, how true this was for him. The *Principles* grew like a living thing in Mendeleev's lifetime, each edition larger, fuller, more mature than its predecessors, each filled with exuberating and spreading footnotes (footnotes which became so enormous that in the last editions they filled more pages than the text; indeed, some occupied nine-tenths of the page—I think my own love of footnotes, the excursions they allow, was partly determined by reading the *Principles*).

Mendeleev, profoundly attached to her, was later to dedicate the *Principles* to her memory.)

Even as a student in St. Petersburg, Mendeleev showed not only an insatiable curiosity, but a hunger for organizing principles of all kinds. Linnaeus, in the eighteenth century, had classified animals and plants, and (much less successfully) minerals, too. Dana, in the 1830s, had replaced the old physical classification of minerals with a chemical classification of a dozen or so main categories (native elements, oxides, sulfides, and so on). But there was no such classification for the elements themselves, and there were now some sixty elements known. Some elements, indeed, seemed almost impossible to categorize. Where did uranium go, or that puzzling, ultralight metal, beryllium? Some of the most recently discovered elements were particularly difficult—thallium, for example, discovered in 1862, was in some ways similar to lead, in others to silver, in others to aluminum, and in yet others to potassium.

It was nearly twenty years from Mendeleev's first interest in classification to the emergence of his periodic table in 1869. This long pondering and incubation (so similar, in a way, to Darwin's before he published *On the Origin of Species*) was perhaps the reason why, when Mendeleev finally published his *Principles,* he could bring a vastness of knowledge and insight far beyond any of his contemporaries—some of them also had a clear vision of periodicity, but none of them could marshal the overwhelming detail he could.

Mendeleev described how he would write the properties and atomic weights of the elements on cards and ponder and shuffle these constantly on his long railway journeys through Russia, playing a sort of patience or (as he called it) "chemical solitaire," groping for an order, a system that might bring sense to all the elements, their properties and atomic weights.

There was another crucial factor. There had been considerable confusion, for decades, about the atomic weights of many elements. It was only when this was cleared up finally, at the Karls-ruhe conference in 1860, that Mendeleev and others could even think of achieving a full taxonomy of the elements. Mendeleev had gone to Karlsruhe with Borodin (this was a musical as well as a chemical journey, for they stopped at many churches en route, trying out the local organs for themselves). With the old, pre-Karlsruhe atomic weights one could get a sense of local triads or groups, but one could not see that there was a numerical rela-tionship *between* the groups themselves.[3] Only when Cannizzaro showed how reliable atomic weights could be obtained and showed, for example, that the proper atomic weights for the alkaline earth metals (calcium, strontium, and barium) were 40, 88, and 137 (not 20, 44, and 68, as formerly believed) did it become clear how close these were to those of the alkali metals — potassium, rubidium, and cesium. It was this closeness, and in turn the closeness of the atomic weights of the halogens —

---

[3] Mendeleev was not the first to see some significance in the atomic weights of elements. When the atomic weights of the alkaline earth metals were estab-lished by Berzelius, Döbereiner was struck by the fact that the atomic weight of strontium was just midway between that of calcium and barium. Was this an accident, as Berzelius thought, or an indication of something important and general? Berzelius himself had just discovered selenium in 1817, and at once realized that (in terms of chemical properties) it "belonged" between sulfur and tellurium. Döbereiner went further, and brought out a quantitative relation-ship too, for its atomic weight was just midway between theirs. And when lithium was discovered later that year (also in Berzelius's kitchen lab), Döbe-reiner observed that it completed another triad, of alkali metals: lithium, sodium, and potassium. Feeling, moreover, that the gap in atomic weight between chlorine and iodine was too great, Döbereiner thought (as Davy had before him) that there must be a third element analogous to them, a halogen, with an atomic weight midway between theirs. (This element, bromine, was discovered a few years later.)

There were mixed reactions to Döbereiner's "triads," with their implication of a correlation between atomic weight and chemical character. Berzelius and Davy were doubtful of the significance of such "numerology," as they saw it; but others were intrigued and wondered whether an obscure but fundamental significance was lurking in Döbereiner's figures.

chlorine, bromine, and iodine—which incited Mendeleev, in 1868, to make a small grid juxtaposing the three groups:

Cl	35.5	K	39	Ca	40
Br	80	Rb	85	Sr	88
I	127	Cs	133	Ba	137

And it was at this point, seeing that arranging the three groups of elements in order of atomic weight produced a repetitive pattern—a halogen followed by an alkali metal, followed by an alkaline earth metal—that Mendeleev, feeling this must be a fragment of a larger pattern, leapt to the idea of a periodicity governing *all* the elements—a Periodic Law.

Mendeleev's first small table had to be filled in, and then extended in all directions, as if filling up a crossword puzzle; this in itself required some bold speculations. What element, he wondered, was chemically allied with the alkaline earth metals, yet followed lithium in atomic weight? No such element apparently existed—or could it be beryllium, usually considered to be trivalent, with an atomic weight of 14.5? What if it was bivalent instead, with an atomic weight, therefore, not of 14.5 but 9? Then it would follow lithium and fit into the vacant space perfectly.

Moving between conscious calculation and hunch, between intuition and analysis, Mendeleev arrived within a few weeks at a tabulation of thirty-odd elements in order of ascending atomic weight, a tabulation that now suggested there was a recapitulation of properties with every eighth element. And on the night of February 16, 1869, it is said, he had a dream in which he saw almost all of the known elements arrayed in a grand table. The following morning, he committed this to paper.[4]

---

[4] This, at least, is the accepted myth, and one that was later promulgated by Mendeleev himself, somewhat as Kekulé was to describe his own discovery of the benzene ring years later, as the result of a dream of snakes biting their own

The logic and pattern of Mendeleev's table were so clear that certain anomalies stood out at once. Certain elements seemed to be in the wrong places, while certain places had no elements. On the basis of his enormous chemical knowledge, he repositioned half a dozen elements, in defiance of their accepted valency and atomic weights. In doing this, he displayed an audacity that shocked some of his contemporaries (Lothar Meyer, for one, felt it was monstrous to change atomic weights simply because they did not "fit").

In an act of supreme confidence, Mendeleev reserved several empty spaces in his table for elements "as yet unknown." He asserted that by extrapolating from the properties of the elements above and below (and also, to some extent, from those to either side) one might make a confident prediction as to what these unknown elements would be like. He did exactly this in his 1871 table, predicting in great detail a new element ("eka-aluminum") which would come below aluminum in Group III. Four years later just such an element was found, by the French chemist Lecoq de Boisbaudran, and named (either patriotically, or in sly reference to himself, *gallus,* the cock) gallium.

The exactness of Mendeleev's prediction was astonishing: he predicted an atomic weight of 68 (Lecoq got 69.9) and a specific gravity of 5.9 (Lecoq got 5.94) and correctly guessed at a great number of gallium's other physical and chemical properties — its fusibility, its oxides, its salts, its valency. There were some initial discrepancies between Lecoq's observations and Mendeleev's predictions, but all of these were rapidly resolved in favor of

---

tails. But if one looks at the actual table that Mendeleev sketched, one can see that it is full of transpositions, crossings-out, and calculations in the margins. It shows, in the most graphic way, the creative struggle for understanding which was going on in his mind. Mendeleev did not wake from his dream with all the answers in place, but, more interestingly, perhaps, woke with a sense of revelation, so that within hours he was able to solve many of the questions that had occupied him for years.

Mendeleev. Indeed, it was said that Mendeleev had a better grasp of the properties of gallium—an element he had never even seen—than the man who actually discovered it.

Suddenly Mendeleev was no longer seen as a mere speculator or dreamer, but as a man who had discovered a basic law of nature, and now the periodic table was transformed from a pretty but unproven scheme to an invaluable guide which could allow a vast amount of previously unconnected chemical information to be coordinated. It could also be used to suggest all sorts of research in the future, including a systematic search for "missing" elements. "Before the promulgation of this law," Mendeleev was to say nearly twenty years later, "chemical elements were mere fragmentary, incidental facts in Nature; there was no special reason to expect the discovery of new elements."

Now, with Mendeleev's periodic table, one could not only expect their discovery, but predict their very properties. Mendeleev made two more equally detailed predictions, and these were also confirmed with the discovery of scandium and germanium a few years later.[5] Here, as with gallium, he made his predictions on the basis of analogy and linearity, guessing that the physical and chemical properties of these unknown elements,

---

[5] In an 1889 footnote—even his lectures had footnotes, at least in their printed versions—he added: "I foresee some more new elements, but not with the same certitude as before." Mendeleev was well aware of the gap between bismuth (with an atomic weight of 209) and thorium (232), and conceived that several elements must exist to fill it. He was most certain of the element immediately following bismuth—"an element analogous to tellurium, which we may call dvi-tellurium." This element, polonium, was discovered by the Curies in 1898, and when finally isolated it had almost all the properties Mendeleev had predicted. (In 1899 Mendeleev visited the Curies in Paris and welcomed radium as his "eka-barium.")

In the final edition of the *Principles,* Mendeleev made many other predictions—including two heavier analogs of manganese—an "eka-manganese" with an atomic weight of around 99, and a "tri-manganese" with an atomic weight of 188; sadly, he never saw these. "Tri-manganese"—rhenium— was not discovered until 1925, the last of the naturally occurring elements

and their atomic weights, would be between those of the neighboring elements in their vertical groups.[6]

The keystone to the whole table, curiously, was not anticipated by Mendeleev, and perhaps could not have been, for this was not a question of a missing element, but of an entire family or group. When argon was discovered in 1894—an element which did not seem to fit anywhere in the table—Mendeleev denied at first that it could be an element and thought it was a heavier form of nitrogen (N_3, analogous to ozone, O_3). But then it became apparent that there *was* a space for it, right between

---

to be found; while "eka-manganese," technetium, was the first new element to be artificially made, in 1937.

He also envisaged, by analogy, some elements following uranium.

[6] It is a remarkable example of synchronicity that in the decade following the Karlsruhe conference there emerged not one but *six* such classifications, all completely independent of one another: de Chancourtois's in France, Odling's and Newlands's, both in England, Lothar Meyer's in Germany, Hinrichs's in America, and finally Mendeleev's in Russia, all pointing toward a periodic law.

De Chancourtois, a French mineralogist, was the first to devise such a classification, and in 1862—just eighteen months after Karlsruhe—he inscribed the symbols of twenty-four elements spiraling around a vertical cylinder at heights proportional to their atomic weights, so that elements with similar properties fell one beneath another. Tellurium occupied the midpoint of the helix; hence he called it a "telluric screw," a *vis tellurique.* But the *Comptes Rendu,* when they came to publish his paper, managed—grotesquely—to omit the crucial illustration, and this, among other problems, put paid to the whole enterprise, causing de Chancourtois's ideas to be ignored.

Newlands, in England, was scarcely any luckier. He, too, arranged the known elements by increasing atomic weight, and seeing that every eighth element, apparently, was analogous to the first, he proposed a "Law of Octaves," saying that "the eighth element, starting from a given one, is a kind of repetition of the first, like the 8th note in an octave of music." (Had the inert gases been known at the time, it would, of course, have been every ninth element that resembled the first.) A too-literal comparison to music, and the suggestion even that these octaves might be a sort of "cosmic music," evoked a sarcastic response at the meeting of the Chemical Society at which Newlands presented his theory; it was said that he might have done as well to arrange the elements alphabetically.

There is no doubt that Newlands, even more than de Chancourtois, was

chlorine and potassium, and indeed, for a whole group coming
between the halogens and the alkali metals in every period. This
was realized by Lecoq, who went on to predict the atomic
weights of the other yet-to-be-discovered gases—and these,
indeed, were discovered in short order. With the discovery of
helium, neon, krypton, and xenon, it was clear that these gases
formed a perfect periodic group, a group so inert, so modest, so
unobtrusive, as to have escaped for a century the chemist's attention.[7] The inert gases were identical in their inability to form
compounds; they had a valency, it seemed, of zero.[8]

---

very close to a periodic law. Like Mendeleev, Newlands had the courage to
invert the order of certain elements when their atomic weight did not match
what seemed to be their proper position in his table (though he failed to make
any predictions of unknown elements, as Mendeleev did).

Lothar Meyer was also at the Karlsruhe conference and was one of the first to
use the revised atomic weights published there in a periodic classification. In
1868 he came up with an elaborate sixteen-columned periodic table (but the
publication of this was delayed until after Mendeleev's table had appeared).
Lothar Meyer paid special attention to the physical properties of the elements
and their relation to atomic weights, and in 1870 he published a famous graph
plotting the atomic weights of the known elements against their "atomic volumes" (this being the ratio of atomic weight to density), a graph that showed
high points for the alkali metals and low points for the dense, small-atomed
metals of Group VIII (the platinum and iron metals), with all the other elements falling nicely in between. This graph proved a most potent argument for
a periodic law and did much to assist the acceptance of Mendeleev's work.

But at the time of discovering his "Natural System," Mendeleev was either
ignorant of, or denied knowledge of, any attempts comparable to his own.
Later, when his name and fame were established, he became more knowledge-
able, perhaps more generous, less threatened by the notion of any codiscoverers
or forerunners. When, in 1889, he was invited to give the Faraday Lecture in
London, he paid a measured tribute to those who had come before him.

[7] Cavendish, however, sparking the nitrogen and oxygen of air together, had
observed in 1785 that a small amount ("not more than ¹⁄₂₀th part of the whole")
was totally resistant to combination, but no one paid any attention to this until
the 1890s.

[8] I think I identified at times with the inert gases, and at other times anthro-
pomorphized them, imagining them lonely, cut off, yearning to bond. Was
bonding, bonding with other elements, absolutely impossible for them? Might

. . .

The periodic table was incredibly beautiful, the most beautiful thing I had ever seen. I could never adequately analyze what I meant here by beauty—simplicity? coherence? rhythm? inevitability? Or perhaps it was the symmetry, the comprehensiveness of every element firmly locked into its place, with no gaps, no exceptions, everything implying everything else.

I was disturbed when one enormously erudite chemist, J. W. Mellor, whose vast treatise on inorganic chemistry I had started dipping into, spoke of the periodic table as "superficial" and "illusory," no truer, no more fundamental than any other ad hoc classification. This threw me into a brief panic, made it imperative for me to see if the idea of periodicity was supported in any ways beyond chemical character and valency.

Exploring this took me away from my lab, took me to a new

---

not fluorine, the most active, the most outrageous of the halogens—so eager to combine that it had defeated efforts to isolate it for more than a century—might not fluorine, if given a chance, at least bond with xenon, the heaviest of the inert gases? I pored over tables of physical constants and decided that such a combination was just, in principle, possible.

In the early 1960s, I was overjoyed to hear (even though my mind at this time had moved on to other things) that the American chemist Neil Bartlett had managed to prepare such a compound—a triple compound of platinum, fluorine, and xenon. Xenon fluorides and xenon oxides were subsequently made.

Freeman Dyson has written to me describing his boyhood love of the periodic table and of the inert gases—he, too, saw them, in their bottles, in the Science Museum in South Kensington—and how excited he was years later when he was shown a specimen of barium xenate, seeing the elusive, unreactive gas firmly and beautifully locked up in a crystal:

> For me too, the periodic table was a passion. . . . As a boy, I stood in front of the display for hours, thinking how wonderful it was that each of these metal foils and jars of gas had its own distinct personality. . . . One of the memorable moments of my life was when Willard Libby came to Princeton with a little jar full of crystals of barium xenate. A stable compound, looking like common salt, but much heavier. This was the magic of chemistry, to see xenon trapped into a crystal.

book that immediately became my bible, the *CRC Handbook of Physics and Chemistry*, a thick, almost cubical book of nearly three thousand pages, containing tables of every imaginable physical and chemical property, many of which, obsessively, I learned by heart.

I learned the densities, melting points, boiling points, refractive indices, solubilities, and crystalline forms of all the elements and hundreds of their compounds. I became consumed with graphing these, plotting atomic weights against every physical property I could think of. I became more and more excited, exuberant, the more I explored, for almost everything I looked at showed periodicity: not only density, melting point, boiling point, but conductivity for heat and electricity, crystalline form, hardness, volume changes with fusion, expansion by heat, electrode potentials, etc., etc. It was not just valency, then, it was physical properties, too. The power, the universality of the periodic table was increased for me by this confirmation.

There were exceptions to the trends shown in the periodic table, anomalies, too—some of them profound. Why, for example, was manganese such a bad conductor of electricity, when the elements on either side of it were reasonably good conductors? Why was strong magnetism confined to the iron metals? And yet these exceptions, I was somehow convinced, reflected special additional mechanisms at work, and in no sense invalidated the overall system.[9]

Using the periodic table, I tried my hand at prediction too, trying to predict the properties of a couple of still-unknown elements as Mendeleev had done for gallium and the others. I had

---

[9] A spectacular anomaly came up with the hydrides of the nonmetals—an ugly bunch, about as inimical to life as one could get. Arsenic and antimony hydrides were very poisonous and smelly; silicon and phosphorus hydrides were spontaneously inflammable. I had made in my lab the hydrides of sulfur

observed, when I first saw the museum table, that there were four gaps in it. The last of the alkali metals, element 87, was still missing, as was the last of the halogens, element 85. Element 43, the one below manganese, was still missing, though this space read "?Masurium" with no atomic weight.[10] Finally there was a rare earth, element 61, missing too.

It was easy to predict the properties of the unknown alkali metal, for the alkali metals were all very similar, and one had only to extrapolate from the other elements in the group. 87, I reckoned, would be the heaviest, must fusible, most reactive of them all; it would be a liquid at room temperature, and like cesium have a golden sheen. Indeed, it might be salmon pink, like molten copper. It would be even more electropositive than cesium, and show an even stronger photoelectric effect. Like the other alkali metals, it would color flames a striking color — prob-

---

(H_2S), selenium (H_2Se), and tellurium (H_2Te), all Group VI elements, all dangerous and vile-smelling gases. The hydride of oxygen, the first Group VI element, one might predict by analogy, would be a foul-smelling, poisonous, inflammable gas, too, condensing to a nasty liquid around $-100°C$. And instead it was water, H_2O — stable, potable, odorless, benign, and with a host of special, indeed unique properties (its expansion when frozen, its great heat capacity, its capacity as an ionizing solvent, etc.) which made it indispensable to our watery planet, indispensable to life itself. What made it such an anomaly? Water's properties did not undermine for me the placement of oxygen in the periodic table, but made me intensely curious as to why it was so different from its analogs. (This question, I found, had only been resolved recently, in the 1930s, with Linus Pauling's delineation of the hydrogen bond.)

[10] Ida Tacke Noddack was one of a team of German scientists who found element 75, rhenium, in 1925–26. Noddack also claimed to have found element 43, which she called masurium. But this claim could not be supported, and she was discredited. In 1934, when Fermi shot neutrons at uranium and thought he had made element 93, Noddack suggested that he was wrong, that he had in fact split the atom. But since she had been discredited with element 43, no one paid any attention to her. Had she been listened to, Germany would probably have had the atomic bomb and the history of the world would have been different. (This story was told by Glenn Seaborg when he was presenting his recollections at a conference in November 1997.)

ably a bluish color, since the flame colors from lithium to cesium tended in this direction.

It was equally easy to predict the properties of the unknown halogen, for the halogens, too, were very similar, and the group showed simple, linear trends.

But predicting the properties of 43 and 61 would be trickier, for these were not "typical" elements (in Mendeleev's term). And it was precisely with such nontypical elements that Mendeleev had run into trouble, leading him to revise his original table. The transition metals had a sort of homogeneity. They were all metals, all thirty of them, and most of them, like iron, were hard and tough, dense and infusible. This was especially so of the heavy transition elements, like the platinum metals and filament metals Uncle Dave had introduced me to. My interest in color brought home another fact, that where compounds of typical elements were usually colorless, like common salt, the compounds of transition metals often had vivid colors: the pink minerals and salts of manganese and cobalt, the green of nickel and copper salts, the many colors of vanadium; going with their many colors were their many valencies, too. All these properties showed me that the transition elements were a special sort of animal, different in nature from the typical elements.

Still, one might hazard a guess that element 43 would have some of the characteristics of manganese and rhenium, the other metals in its group (it would, for instance, have a maximum valency of 7, and form colored salts); but it would also be generically similar to the neighboring transition metals in its period—niobium and molybdenum to the left, and the light platinum metals to the right. So one could also predict that it would be a shining, hard, silvery metal with a density and melting point similar to theirs. It would be just the sort of metal Uncle Tungsten would love, and just the sort of metal which would have been discovered by Scheele in the 1770s—that is, if it existed in sensible amounts.

The hardest prediction, in any detail, would be for element 61, the missing rare earth metal, for these elements were in many ways the most baffling of all.

I think I first heard of the rare earths from my mother, who was a chain smoker and lit cigarette after cigarette with a small Ronson lighter. She showed me the "flint" one day, pulling it out, and said it was not really flint, but a metal that produced sparks when it was scratched. This "mischmetal"—cerium mostly—was a mishmash of half a dozen different metals, all of them very similar, all of them rare earths. This odd name, the rare earths, had a mythical or fairy-tale sound to it, and I imagined the rare earths as not only rare and precious, but as having special, secret qualities possessed by nothing else.

Later Uncle Dave told me of the extraordinary difficulty which chemists had had in separating the individual rare earths—there were a dozen or more—for they were astoundingly similar, at times indistinguishable in their physical and chemical properties. Their ores (which for some reason all seemed to come from Sweden) never contained a single rare-earth element, but a whole cluster of them, as if nature herself had trouble distinguishing them. Their analysis formed a whole saga in chemical history, a saga of passionate research (and frequently frustration) in the hundred years or more it took to identify them. The separation of the last few rare-earth elements, indeed, was beyond the powers of chemistry in the nineteenth century, and it was only with the use of physical methods such as spectroscopy and fractional crystallization that they were finally separated. No fewer than fifteen thousand fractional crystallizations, exploiting the infinitesimal differences in solubility between their salts, were needed to separate the final two, ytterbium and lutecium—an enterprise that occupied years.

Nonetheless there were chemists who were enthralled with the intransigent rare-earth elements and spent their entire lives trying to isolate them, sensing that their study might cast an unexpected light on all the elements and their periodicities:

> The rare earths [wrote William Crookes] perplex us in our researches, baffle us in our speculations, and haunt us in our very dreams. They stretch like an unknown sea before us, mocking, mystifying, and murmuring strange revelations and possibilities.

If the rare-earth elements baffled, mocked, and haunted chemists, they positively maddened Mendeleev as he struggled to assign them a place in his periodic table. There were only five rare earths known when he constructed his first table in 1869, but then more and more were discovered in the decades that followed, and with each discovery the problem grew, because all of them, with their consecutive atomic weights, belonged (it seemed) in a single space in the table, crushed, as it were, between two adjoining elements in Period 6. Others, too, struggled with the placement of the maddeningly similar elements, further frustrated by a deep uncertainty as to how many rare-earth elements there might ultimately prove to be.

Many chemists, by the end of the nineteenth century, were inclined to put both the transition and the rare-earth elements into separate "blocks," for one needed a periodic table with more space, more dimensions, to accommodate these "extra" elements that seemed to interrupt the basic eight groups of the table. I tried making different forms of periodic table myself to accommodate these blocks, experimenting with spiral ones and three-dimensional ones. Many others, I later found, had done the same: more than a hundred versions of the table appeared during Mendeleev's lifetime.

All of the tables I made, all of the tables I saw, ended with uncertainty, ended with a question mark, centered around the "last" element, uranium. I was intensely curious about this, about Period 7, which started with the as-yet-unknown alkali metal, element 87, but only got as far as uranium, element 92. Why, I wondered, should it stop here, after only six elements? Could there not be more elements, beyond uranium?

Uranium itself had been placed by Mendeleev under tungsten, the heaviest of the Group VI transition elements, for it was very much like tungsten, chemically. (Tungsten formed a volatile hexafluoride, a very dense vapor, and so did uranium—this compound, UF_6, was used in the war to separate out the isotopes of uranium.) Uranium *seemed* like a transition metal, *seemed* like eka-tungsten—and yet, I felt somehow uncomfortable about this, and decided to do a little exploring, to examine the densities and melting points of all the transition metals. As soon as I did this I discovered an anomaly, for where the densities of the metals steadily increased through Periods 4, 5, and 6, they unexpectedly declined when one came to the elements in Period 7. Uranium was actually *less* dense than tungsten, though one would have expected it to be more so (thorium, similarly, was less dense than hafnium, not more so, as one would have expected). It was precisely the same with their melting points: these reached a maximum in Period 6, then suddenly declined.

I was excited about this; I felt I had made a discovery. Was it possible, despite all the similarities between uranium and tungsten, that uranium did *not* in fact belong in the same group, was not even a transition metal at all? Might this also be the case for the other Period 7 elements, thorium and protoactinium, and the (imaginary) elements beyond uranium? Could it be that these elements were instead the beginning of a second rare-earth series precisely analogous to the first one in Period 6? If this was the case, then eka-tungsten would not be uranium, but an as-yet-undiscovered element, which would appear only after the second rare-earth series had completed itself. In 1945, this was still unimaginable, the stuff of science fiction.

. . .

I was thrilled, soon after the war, to find that I had guessed right, when it was revealed that Glenn Seaborg and his coworkers in Berkeley had succeeded in making a number of transuranic elements—elements 93, 94, 95, and 96—and found that these indeed were part of a second series of rare-earth elements (which, by analogy with the first rare-earth series, the lanthanides, he called the actinides).[11]

The number of elements in the second series of rare earths, Seaborg argued, by analogy with the first series, would also be fourteen, and after the fourteenth (element 103) one might expect ten transition elements, and only then the concluding elements of Period 7, ending with an inert gas at element 118. Beyond this, Seaborg suggested, a new period would start, beginning, like all the others, with an alkali metal, element 119.

It seemed that the periodic table might thus be extended to new elements far beyond uranium, elements that might not even exist in nature. Whether there was any limit to such transuranic elements was not clear: perhaps the atoms of such elements would become too big to hold together. But the principle of periodicity was fundamental, and could be extended, it seemed, indefinitely.

While Mendeleev saw the periodic table primarily as a tool for organizing and predicting the properties of the elements, he also felt it embodied a fundamental law, and he wondered on occasion

---

[11] Although elements 93 and 94, neptunium and plutonium, were created in 1940, their existence was not made public until after the war. They were given provisional names, when they were first made, of "extremium" and "ultimium," because it was thought impossible that any heavier elements would ever be made. Elements 95 and 96, however, were created in 1944. Their discovery was not made public in the usual way—in a letter to *Nature*, or at a meeting of the Chemical Society—but during a children's radio quiz show in November 1945, during which a twelve-year-old boy asked, "Mr. Seaborg, have you made any more elements lately?"

about "the invisible world of chemical atoms." For the periodic table, it was clear, looked both ways: outward to the manifest properties of the elements, and inward to some as-yet-unknown atomic property which determined these.

In that first, long, rapt encounter in the Science Museum, I was convinced that the periodic table was neither arbitrary nor superficial, but a representation of truths which would never be overturned, but would, on the contrary, continually be confirmed, show new depths with new knowledge, because it was as deep and simple as nature itself. And the perception of this produced in my twelve-year-old self a sort of ecstasy, the sense (in Einstein's words) that "a corner of the great veil had been lifted."

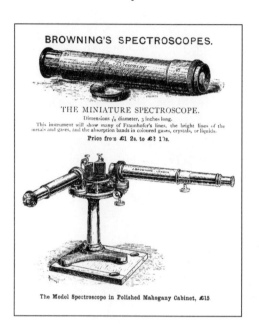

17

BROWNING'S SPECTROSCOPES.

THE MINIATURE SPECTROSCOPE.

Dimensions ⅝ diameter, 3 inches long.

This instrument will show many of Fraunhofer's lines, the bright lines of the metals and gases, and the absorption bands in coloured gases, crystals, or liquids.

Price from £1 2s. to £2 1)s.

The Model Spectroscope in Polished Mahogany Cabinet, £15.

A POCKET SPECTROSCOPE

We had always celebrated Guy Fawkes night, before the war, by setting off fireworks. Bengal lights, burning brilliantly green or red, were my favorites. The green, my mother had told me, was due to an element called barium, the red to strontium. I had no idea at that point what barium and strontium were, but their names, like their colors, stayed in my mind.

When my mother saw how enthralled I was by these lights, she showed me how, if one threw a pinch of salt on the stove, the gas flame suddenly flared and turned a brilliant yellow—this was due to the presence of another element, sodium (even the Romans, she said, had used it to give their fires and flares a richer color). So, in a sense, I was introduced to "flame tests" even before the war, but it was only a few years later, in Uncle Dave's lab, that I learned they were an essential part of chemical life, an instant way of detecting certain elements, even if present in minute amounts.

One had only to put a speck of the element or one of its compounds on a loop of platinum wire and put this in the colorless flame of a Bunsen burner to see the colorations produced. I explored a whole range of flame colors. There was the azure blue flame produced by copper chloride. And there was the light blue—the "poisonous" light blue, as I regarded it—produced by lead and arsenic and selenium. There were lots of green flames: an emerald green with most other copper compounds; a yellowish green with barium compounds, some boron compounds too—borane, boron hydride, was highly inflammable and burned with an eerie green flame of its own. Then there were the red ones: the carmine flame of lithium compounds, the scarlet of strontium, the yellowish brick red of calcium. (I read later that radium also colored flames red, but this, of course, I was never to see. I imagined it as a red of the most refulgent brilliance, a sort of ultimate, fatal red. The chemist who first saw it, so I imagined, went blind soon after, the radioactive, retina-destroying red of radium being the last thing he ever saw.)

These flame tests were very sensitive—much more so than many chemical reactions, the "wet" tests one also did to analyze substances—and they reinforced a sense of elements as fundamental, as retaining their unique properties however they were combined. Sodium, one might feel, was "lost" when it combined with chlorine to form salt—but the telltale presence of sodium yellow in a flame test served to remind one that it was still there.

Auntie Len had given me James Jeans's book *The Stars in Their*

Courses for my tenth birthday, and I had been intoxicated by the imaginary journey Jeans described into the heart of the sun, and his casual mention that the sun contained platinum and silver and lead, most of the elements we have on earth.

When I mentioned this to Uncle Abe, he decided it was time for me to learn about spectroscopy. He gave me an 1873 book, *The Spectroscope,* by J. Norman Lockyer, and lent me a small spectroscope of his own. Lockyer's book had charming illustrations showing not just various spectroscopes and spectra, but bearded, frock-coated Victorian scientists examining candle flames with the new apparatus, and it gave me a very personal sense of the history of spectroscopy, from Newton's first experiments to Lockyer's own pioneering observations of the spectra of the sun and stars.

Spectroscopy indeed had started in the heavens, with Newton's decomposition of sunlight with a prism in 1666, showing that it was composed of rays "differently refrangible." Newton obtained the sun's spectrum as a continuous luminous band of color going from red to violet, like a rainbow. A hundred and fifty years later, Joseph Fraunhofer, a young German optician, using a much finer prism and a narrow slit, was able to see that the entire length of Newton's spectrum was interrupted by odd dark lines, "an infinite number of vertical lines of different thicknesses" (he was able, finally, to count more than five hundred).

One needed a brilliant light to get a spectrum, but it did not have to be sunlight. It could be the light of a candle, or limelight, or the colored flames of the alkali or alkaline earth metals. By the 1830s and 1840s these, too, were being examined, and an entirely different sort of spectrum was now seen. Whereas sunlight produced a luminous band with every spectral color in it, the light of vaporized sodium produced only a single yellow line, a very narrow line of great brilliance, set upon a background of inky blackness. It was similar with the flame spectra of lithium and strontium, except these had a multitude of bright lines, mostly in the red part of the spectrum.

What was the origin of the dark lines Fraunhofer saw in 1814? Had they any relation to the bright spectral lines of flamed elements? These questions presented themselves to many minds at the time, but remained unanswered until 1859, when Gustav Kirchhoff, a young German physicist, joined forces with Robert Bunsen. Bunsen was a distinguished chemist by this time, and a prolific inventor—he had invented photometers, calorimeters, the carbon-zinc cell (still used, with negligible change, in the batteries I pulled to pieces in the 1940s), and, of course, the Bunsen burner, which he had perfected to investigate color phenomena more closely. They were an ideal pair, Bunsen a superb experimentalist—practical, technically brilliant, inventive—and Kirchhoff with a theorizing power, a mathematical facility, that Bunsen perhaps lacked.

In 1859, Kirchhoff performed a simple and beautifully designed experiment, which showed that the bright-line and dark-line spectra—the emission and the absorption spectra—were one and the same, the corresponding opposites of the same phenomenon: the capacity of elements to emit light of characteristic wavelength when vaporized, or to absorb light of exactly the same wavelength if they were illuminated. Thus the characteristic line of sodium could be seen either as a brilliant yellow line in its emission spectrum, or as a dark line in exactly the same position in its absorption spectrum.

Directing his spectroscope to the sun, Kirchhoff realized that one of the countless dark Fraunhofer lines in the solar spectrum was in exactly the same position as the bright yellow line of sodium—and that the sun, therefore, must contain sodium. The general feeling, in the first half of the nineteenth century, had been that we would never know anything about the stars beyond what could be gained by simple observation—that their composition and chemistry, in particular, would remain perpetually unknown, and so Kirchhoff's discovery was greeted with astonishment.[1]

Kirchhoff and others (and especially Lockyer himself) went on to identify a score of other terrestrial elements in the sun, and now the Fraunhofer mystery—the hundreds of black lines in the solar spectrum—could be understood as the absorption spectra of these elements in the outermost layers of the sun, as they were transilluminated from within. On the other hand, a solar eclipse, it was predicted, with the central brilliance of the sun obscured and only its brilliant corona visible, would produce instead dazzling emission spectra corresponding to the dark lines.

Now, with Uncle Abe's help—he had a small observatory on the roof of his house, and kept one of his telescopes hitched up to a spectroscope—I saw this for myself. The whole visible universe—planets, stars, distant galaxies—presented itself for spectroscopic analysis, and I got a vertiginous, almost ecstatic satisfaction from seeing familiar terrestrial elements out in space, seeing what I had known only intellectually before, that the elements were not just terrestrial but cosmic, were indeed the building blocks of the universe.

At this point, Bunsen and Kirchhoff turned their attention away from the heavens, to see if they could find any new or undiscovered elements on the earth using their new technique. Bunsen had already observed the great power of the spectroscope to resolve complex mixtures—to provide, in effect, an optical analysis of chemical compounds. If lithium, for example, was present in small amounts along with sodium, there was no way,

---

[1] Auguste Comte had written, in his 1835 *Cours de la Philosophie Positive:*

On the subject of the stars, all investigations which are not ultimately reducible to simple visual observations are . . . necessarily denied to us. While we can conceive of the possibility of determining their shapes, their sizes, and their motions, we shall never be able by any means to study their chemical composition or mineralogical content.

with conventional chemical analysis, to detect it. Nor were flame colors of help here, because the brilliant yellow flame of sodium tended to flood out other flame colors. But with a spectroscope, the characteristic spectrum of lithium could be seen immediately, even if it was mixed with ten thousand times its weight of sodium.

This enabled Bunsen to show that certain mineral waters rich in sodium and potassium also contained lithium (this had been completely unsuspected, the only sources hitherto having been certain rare minerals). Could they contain other alkali metals too? When Bunsen concentrated his mineral water, rendering down 600 quintals (about 44 tons) to a few liters, he saw, amid the lines of many other elements, two remarkable blue lines, close together, which had never been seen before. This, he felt, must be the signature of a new element. "I shall name it cesium because of its beautiful blue spectral line," he wrote, announcing its discovery in November 1860.

Three months later, Bunsen and Kirchhoff discovered another new alkali metal; they called this rubidium, from "the magnificent dark red color of its rays."

Within a few decades of Bunsen and Kirchhoff's discoveries twenty more elements were discovered with the aid of spectroscopy—indium and thallium (which were also named for their brilliantly colored spectral lines), gallium, scandium, and germanium (the three elements Mendeleev had predicted), all the remaining rare-earth elements, and, in the 1890s, the inert gases.

But perhaps the most romantic story of all, certainly the one that most appealed to me as a boy, had to do with the discovery of helium. It was Lockyer himself who, during a solar eclipse in 1868, was able to see a brilliant yellow line in the sun's corona, a line near the yellow sodium lines, but clearly distinct from them. He surmised that this new line must belong to an element unknown on earth, and named it helium (he gave it the metallic suffix of -*ium* because he assumed it was a metal). This finding aroused great wonder and excitement, and it was even speculated by some that every star might have its own special elements. It was only twenty-five years later that certain terrestrial (uranium) minerals were found to contain a strange, light gas, readily released, and when this was submitted to spectroscopy it proved to be the selfsame helium.

The wonder of spectral analysis, analysis at a distance, had literary resonances as well. I had read *Our Mutual Friend* (written in 1864, just four years after Bunsen and Kirchhoff had launched spectroscopy), and here Dickens imagined a "moral spectroscopy" whereby the inhabitants of remote galaxies and stars might analyze the light from the Earth to gauge its good and evil, the moral spectrum of its inhabitants.

"I have little doubt," Lockyer wrote at the end of his book, "that, as time rolls on . . . the spectroscope [will] become . . . the pocket companion of everyone amongst us." A small spectroscope became my own constant companion, my instant analyzer of the world, whipped out on all sorts of occasions: to look at the new fluorescent lights that were beginning to appear in London Tube stations, to look at solutions and flames in my lab, or at coal fires and gas flames in the house.

I also explored the absorption spectra of compounds of all sorts, from simple inorganic solutions to blood, leaves, urine, and wine. I was fascinated to find out how characteristic the spectrum of blood was even when dried and how small a quantity was needed to analyze in this fashion—one could identify a faint bloodstain more than fifty years old and distinguish it from a rust stain. The forensic possibilities of this intrigued me; I wondered if Sherlock Holmes, along with his chemical explorations, had used a spectroscope too. (I was especially fond of the Sherlock Holmes stories, and even more of the Professor Challenger ones which Conan Doyle had written later—I identified with Challenger; I could not identify with Holmes. In *The Poison Belt,* spectroscopy plays a crucial role, for it is a change in the Fraunhofer lines of the sun's spectrum that alerts Challenger to the presence of an approaching poison cloud.)

But it was the bright lines, the brilliant colors, the emission spectra I always came back to. I remember going to Piccadilly Circus and Leicester Square with my pocket spectroscope, and looking at the new sodium lights that were being used for street lighting, at the scarlet neon advertisements, and at the other gas-discharge tubes—yellow, blue, green, according to the gas used—which now turned the West End into a glory of colored lights after the long blackout of the war. Each gas, each substance, had its own unique spectrum, its own signature.

Bunsen and Kirchhoff had felt that the position of the spectral lines was not only a unique signature of each element, but a manifestation of its ultimate nature. They seemed to be "a property of a similar unchangeable and fundamental nature as the atomic weight," indeed a manifestation—as yet hieroglyphic and indecipherable—of their very constitution.

The complexity of spectra (that of iron, for example, contained several hundred lines) in itself suggested that atoms could hardly be the small, dense masses which Dalton had imagined, distinguished by their atomic weights and little else.

One chemist, W. K. Clifford, writing in 1870, expressed this complexity in terms of a musical metaphor:

> . . . a grand piano must be a very simple mechanism compared with an atom of iron. For in the spectrum of iron there is an almost innumerable wealth of separate bright lines, each one of which corresponds to a sharp definite period of vibration of the iron atom. Instead of the hundred-odd sound vibrations which a grand piano can emit, the single iron atom appears to emit thousands of definite light vibrations.

There were a variety of such musical images and metaphors at the time, all concerned with the ratios, the harmonics, which seemed to lurk in the spectra, and the possibility of expressing them in a formula. The nature of these "harmonics" remained unclear until 1885, when Balmer was able to find a formula relating the position of the four lines in the visible spectrum of hydrogen, a formula that enabled him to predict correctly the existence and position of further lines in the ultraviolet and infrared. Balmer, too, thought in musical terms, and wondered whether it might be "possible to interpret the vibrations of the individual spectral lines as overtones of, so to say, one specific keynote." That Balmer was on to something of fundamental importance, and not some numerological mumbo jumbo, was immediately recognized, but the implications of his formula were wholly enigmatic—as enigmatic as Kirchhoff's discovery that the emission and absorption lines of elements were the same.

18

COLD FIRE

My many uncles and aunts and cousins served as a sort of archive or reference library, and I would be referred to different ones for specific problems: most often to Auntie Len, my botanical aunt, who had played such a lifesaving role in the grim days of Braefield, or Uncle Dave, my chemical and mineralogical uncle, but there was also Uncle Abe, my physics uncle, who had started me on spectroscopy. Uncle Abe was consulted rather rarely at first, because he was one of the senior uncles, six years Uncle Dave's senior and fifteen years my mother's. He was regarded as the most brilliant of his father's eighteen children. He was intellectually formidable, although his knowledge had come through a sort of osmosis, not formal training. Like

Dave, he had grown up with a taste for physical science, and like Dave, he had gone geologizing to South Africa as a young man.

The great discoveries of X-rays, radioactivity, the electron, and quantum theory had all occurred in his formative years and were to remain central interests for the rest of his life; he had a passion for astronomy and for number theory as well. But he was also perfectly capable of turning his mind to practical and commercial ends too. He played a part in developing Marmite, the widely used vitamin-rich yeast extract developed early in the century (my mother adored this; I hated it), and, in the Second World War, when normal soap was difficult to get, he helped develop an effective fat-free soap.

Though Abe and Dave were alike in some ways (both had the broad Landau face, with wide-set eyes, and the unmistakable, resonant Landau voice—characteristics still marked in the great-great-grandchildren of my grandfather), they were very different in others. Dave was tall and strong, with a military posture (he had served in the Great War and in the Boer War before that), always carefully dressed. He would wear a wing collar and highly polished shoes even when he worked at his lab bench. Abe was smaller, somewhat gnarled and bent (in the years that I knew him), brown and grizzled, like an old shikari, with a hoarse voice and chronic cough; he cared little what he wore, and usually had on a sort of rumpled lab smock.

The two were associated formally as codirectors of Tungstalite, though Abe left the business end to Dave and spent all his time in research. It was he who developed a safe and effective way of "pearling" lightbulbs with hydrofluoric acid in the early 1920s—he had designed the machines to do this in the Hoxton factory. He also worked on the use of "getters" in vacuum tubes—highly reactive, oxygen-hungry metals like cesium and barium which could remove the last traces of air from a tube—and, earlier, he had patented the use of Hertzite, his synthetic crystal, for crystal radios.

He had developed and patented a luminous paint, and this was used in military gunsights in the First World War (it may have been decisive, he told me, in the Battle of Jutland). His paints were also used to illuminate the dials of Ingersoll watches and clocks. He had, like Uncle Dave, big, capable hands, but where Uncle Dave's were seamed with tungsten, Uncle Abe's were covered with radium burns and malignant warts from his long, careless handling of radioactive materials.

Both Uncle Dave and Uncle Abe were intensely interested in light and lighting, as was their father; but with Dave it was "hot" light, and with Abe "cold" light. Uncle Dave had drawn me into the history of incandescence, of the rare earths and metallic filaments which glowed and incandesced brilliantly when heated. He had inducted me into the energetics of chemical reactions — how heat was absorbed or emitted during the course of these; heat that sometimes became visible as fire and flame.

Through Uncle Abe, I was drawn into the history of "cold" light — luminescence — which started perhaps before there was any language to record things, with observations of fireflies and glowworms and phosphorescent seas; of will-o'-the-wisps, those strange, wandering, faint globes of light that would, in legend, lure travelers to their doom. And of Saint Elmo's fire, the eerie luminous discharges that could stream in stormy weather from a ship's masts, giving its sailors a feeling of bewitchment. There were the auroras, the Northern and Southern Lights, with their curtains of color shimmering high in the sky. A sense of the uncanny, the mysterious, seemed to inhere in these phenomena of cold light — as opposed to the comforting familiarity of fire and warm light.

There was even an element, phosphorus, which glowed spontaneously. Phosphorus attracted me strangely, dangerously, because of its luminosity — I would sometimes slip down to my lab at night to experiment with it. As soon as I had my fume cupboard set up, I put a piece of white phosphorus in water and boiled it, dimming the lights so that I could see the steam coming out of the flask, glowing a soft greenish blue. Another, rather beautiful experiment was boiling phosphorus with caustic potash in a retort—I was remarkably nonchalant about boiling up such virulent substances—and this produced phosphoretted hydrogen (the old term), or phosphine. As the bubbles of phosphine escaped, they took fire spontaneously, forming beautiful rings of white smoke.

I could ignite phosphorus in a bell jar (using a magnifying glass), and the jar would fill with a "snow" of phosphorus pentoxide. If one did this over water, the pentoxide would hiss, like red-hot iron, as it hit the water and dissolved, making phosphoric acid. Or by heating white phosphorus, I could transform it into its allotrope, red phosphorus, the phosphorus of matchboxes.[1] I had learned as a small child that diamond and graphite were different forms, allotropes, of the same element. Now, in the lab, I could effect some of these changes for myself, turning white phosphorus into red phosphorus, and then (by condensing its vapor) back again. These transformations made me feel like a magician.[2]

But it was especially the luminosity of phosphorus that drew

---

[1] Uncle Abe told me something of the history of matches, how the first matches had to be dipped into sulfuric acid to light them before "lucifers"—friction matches—were introduced in the 1830s, and how this led to a huge demand for white phosphorus over the next century. He told me of the awful conditions under which match girls worked in the factories and of the terrible disease, "phossy-jaw," they often got, until the use of white phosphorus was banned in 1906. (Only red phosphorus, far more stable, and far safer, was subsequently used.)

Abe also spoke of the hellish phosphorus bombs used in the Great War, and how there was a move to ban these, as poison gas had been banned. But now, in 1943, they were being used freely once again, and thousands of people on both sides were being burned alive in the most agonizing way possible.

[2] Phosphorus, oxidizing slowly, was not the only element to glow when

me again and again. One could easily dissolve some of it in clove oil or cinnamon oil, or in alcohol (as Boyle had done)—this not only overcame its garlicky smell, but allowed one to experiment with its luminosity safely, for such a solution might contain only one part of phosphorus in a million, and yet still glow. One could rub a bit of this solution on one's face or hands, and they would glow, ghostlike, in the dark. This glow was not uniform, but seemed (as Boyle had put it) "to tremble much, and sometimes . . . to blaze out with sudden flashes."

Hennig Brandt of Hamburg had been the first to obtain this marvelous element, in 1669. He distilled it from urine (apparently with some alchemical ambition in mind), and he adored the strange, luminous substance he had isolated, and called it cold fire (*kaltes Feuer*), or, in a more affectionate mood, *mein Feuer.*

Brandt handled his new element rather carelessly, and was apparently surprised to discover its lethal powers, as he wrote in a letter to Leibniz on April 30, 1679:

> When in these days I had some of that very fire in my hand and did nothing more than blow on it with my breath, the fire ignited itself as God is my witness; the skin on my hand was burned truly into a hardened stone such that my children cried and declared that it was horrible to witness.

But though all the early researchers burned themselves severely with phosphorus, they also saw it as a magical substance that seemed to carry within itself the radiance of glowworms, perhaps of the moon, a secret, inexplicable radiance of its own. Leibniz, corresponding with Brandt, wondered whether the glowing light of phosphorus could be used for lighting rooms at

---

exposed to air. Sodium and potassium did this too, when they were freshly cut, but lost their luminosity in a few minutes as the cut surfaces tarnished. I found this by chance as I was working in my lab late one afternoon, as it gradually darkened into dusk—I had not yet switched on the light.

night (this, Abe told me, was perhaps the first suggestion of using cold light for illumination).

No one was more intrigued by this than Boyle, who made detailed observations of its luminescence—how it, too, required the presence of air, how it fluctuated strangely. Boyle had already made extensive investigations of "luciferous" phenomena, from glowworms to luminous wood and tainted meat, and had made careful comparisons of such "cold" light with that of glowing coals (both, he found, needed air to sustain them).

On one occasion Boyle was called down from his bedchamber by his frightened and astonished servant, who reported that some meat was glowing brightly in the dark pantry. Boyle, fascinated, got up at once and commenced an investigation, which culminated in his charming paper "Some Observations about Shining Flesh, both of Veal and Pullet, and that without any sensible Putrefaction in those Bodies." (The shining was probably due to luminescent bacteria, but no such organism was known or suspected in Boyle's time.)

Uncle Abe, too, was fascinated by such chemical luminescence, and had experimented a great deal with it as a young man, and with luciferins, the light-producing chemicals of luminous animals. He had wondered whether they could be turned to practical use, to make a really brilliant luminous paint. Chemical luminosity could indeed be dazzlingly brilliant; the only problem was that it was ephemeral, transient, by nature, disappearing as soon as the reactants were consumed—unless there could be (as with fireflies) a continued production of the luciferous chemicals. If chemistry was not the answer, then one needed some other form of energy, something that could be transformed into visible light.

Abe's interest in luminescence had been stimulated, when he was growing up, by a luminous paint used in their old house in Leman Street—Balmain's Luminous Paint, it was called—for painting keyholes, gas and electric fixtures, anything that had to be located in the dark. Abe found these glowing keyholes and switches wonderful, the way they glowed softly for hours after being exposed to light. This sort of phosphorescence had been discovered in the seventeenth century, by a shoemaker in Bologna who had gathered some pebbles, roasted them with charcoal, and then observed that they glowed in the dark for hours after they had been exposed to daylight. This "phosphorus of Bologna," as it was called, was barium sulfide, produced by the reduction of the mineral barytes. Calcium sulfide was easier to procure—it could be made by heating oyster shells with sulfur—and this, "doped" with various metals, was the basis of Balmain's Luminous Paint. (These metals, Abe told me, added in minute amounts, "activated" the calcium sulfide, and lent it different colors as well. Perfectly pure calcium sulfide, paradoxically, did not glow.)

While some substances emitted light slowly in the dark after being exposed to daylight, others glowed only while they were being illuminated. This was fluorescence (after the mineral fluorite, which often showed it). This strange luminosity had been originally discovered as early as the sixteenth century, when it was found that if a slanting beam of light was directed through tinctures of certain woods, a shimmering color might appear in its path—Newton had attributed this to "internal reflection." My father liked to demonstrate it with quinine water—tonic—which showed a faint blue in daylight and a brilliant turquoise in ultraviolet light. But whether a substance was fluorescent or phosphorescent (many were both), it required blue or violet light or daylight (which was rich in light of all wavelengths) to elicit the luminescence—red light was of no use whatever. The most effective illumination, indeed, was invisible—the ultraviolet light that lay beyond the violet end of the spectrum.

My own first experiences of fluorescence occurred with the ultraviolet lamp my father kept in the surgery—an old mercury vapor lamp with a metal reflector, which emitted a dim bluish violet light and an invisible blaze of ultraviolet. It was used to diagnose some skin diseases (certain fungi fluoresced in its light) and to treat others—though my brothers also used it for tanning.

These invisible ultraviolet rays were quite dangerous—one could be severely burned if exposed too long, and one had to wear special goggles like an aviator's, all leather and wool, with thick lenses made of a special glass that blocked most of the ultraviolet (much of the visible, too). Even with goggles, one had to avoid looking directly at the lamp, otherwise a strange, unfocused glow appeared, due to the fluorescence of one's eyeballs. One could see, looking at other people in the ultraviolet light, how their teeth and eyes glowed a brilliant white.

Uncle Abe's house, a short walk from ours, was a magical place, filled with all sorts of apparatus: Geissler tubes, electro-magnets, electric machines and motors, batteries, dynamos, coils of wire, X-ray tubes, Geiger counters and phosphorescent screens, and a variety of telescopes, many of which he had built with his own hands. He would take me up to his attic laboratory, on weekends especially, and once he had satisfied himself that I could handle the apparatus, he gave me the run of his phosphors and fluorescent materials, as well as the little handheld Wood's UV lamp he used (this was much easier to deal with than the old mercury vapor UV lamp we had at home).

Abe had racks and racks of phosphors in his attic, which he would blend like an artist with his palette—the deep blue of cal-cium tungstate, the paler blue of magnesium tungstate, the red of yttrium compounds. Like phosphorescence, fluorescence could often be induced by "doping," adding activators of various sorts, and this was one of Abe's chief research interests, for fluorescent lights were just beginning to come into their own, and subtle phosphors were needed to produce a visible light that was soft and warm and agreeable to the eye.[3] Abe was especially attracted to the very pure and delicate colors which could be made if one added various rare earths as activators—europia, erbia, terbia. Their presence in certain minerals, he told me, even in minute quantities, lent these minerals their special fluorescence.

But there were also substances that would fluoresce even when absolutely pure, and here uranium salts (or, properly speaking, uranyl salts) were preeminent. Even if one dissolved uranyl salts in water, the solutions would be fluorescent—one part in a million was sufficient. The fluorescence could also be transferred to glass, and uranium glass or "canary glass" had been very popular in Victorian and Edwardian houses (it was this which so fascinated me in the stained glass in our front door). Canary glass transmitted yellow light and was usually yellow to look through, but fluoresced a brilliant emerald green under the impact of the shorter wavelengths in daylight, so it would often appear to shimmer, shifting between green and yellow depending on the angle of illumination. And though the stained glass in our front door had been shattered by a bomb blast during the Blitz (it was replaced by an unpleasant, bobbly white glass), its colors, intensified by nostalgia perhaps, still remained preternaturally vivid

---

[3] Equally important were cathode-ray tubes, which were now being developed for television. Abe himself had one of the original television sets of the 1930s, a huge, bulky thing with a tiny circular screen. Its tube, he said, was not much different from the cathode-ray tubes that Crookes had developed in the 1870s, except that its face was coated with a suitable phosphor. Cathode-ray tubes in use for medical or electronic apparatus were often coated with zinc silicate, willemite, which emitted a brilliant green light when bombarded, but for television one needed phosphors that would give a clear, white light—and if color television was to be developed, one would need three separate phosphors with exactly the right balance of color emissions, like the three pigments in color photography. The old dopants used in luminous paints were quite unsuitable for this; much more delicate and precise colors were needed.

in my memory—especially now that Uncle Abe had explained its secret to me.[4]

Though Abe had expended much effort on the development of luminous paints, and later on phosphors for cathode-ray tubes, his central interest, like Dave's, was in the challenge of illumination. The hope he had nourished, from early on, was that it might be possible to develop a form of cold light as efficient, as pleasant, as tractable, as hot light. Thus while Uncle Tungsten's thoughts were fixed on incandescence, it was clear to Uncle Abe from the start that no really powerful cold light could be made without electricity, and that electroluminescence would have to be the key.

That rarefied gases and vapors glowed when electrically charged had been known since the seventeenth century, when it was observed that the mercury in a barometer could become electrified by friction against the glass, and this would set up a beautiful bluish glow in the rarefied mercury vapor in the near vacuum above.[5]

---

[4] Uncle Abe also showed me other types of cold light. One could take various crystals—like uranyl nitrate crystals, or even ordinary cane sugar—and crush them, with a mortar and pestle, or between two test tubes (or even one's teeth), cracking the crystals against one another—this would cause them to glow. This phenomenon, called triboluminescence, was recognized even in the eighteenth century, when Father Giambattista Beccaria recorded:

> You may, when in the dark frighten simple people only by chewing lumps of sugar, and, in the meantime, keeping your mouth open, which will appear to them as if full of fire; to this add, that the light from sugar is the more copious in proportion as the sugar is purer.

Even crystallization could cause luminescence; Abe suggested that I make a saturated solution of strontium bromate and then let it cool slowly in the dark—at first nothing happened, and then I began to see scintillations, little flashes of light, as jagged crystals formed on the bottom of the flask.

[5] The same phenomenon, I read, had been used ingeniously to make self-luminous buoys—these were encircled by rings of strong glass tubing containing mercury under reduced pressure, which would be swirled against the glass and electrified by the motion of the waves.

Using the powerful discharges from the induction coils invented in the 1850s, it was found that a long column of mercury vapor could be set glowing (Alexandre-Edmond Becquerel suggested, early on, that coating the discharge tube with a fluorescent substance might make it more suitable for illumination). But when mercury-vapor lamps were introduced, for special purposes, in 1901, they were dangerous and unreliable, and their light—in the absence of a fluorescent coating—was too blue to allow domestic use. Attempts to coat such tubes with fluorescent powders before the First World War collapsed before a multitude of problems. Other gases and vapors, meanwhile, were being tried: carbon dioxide gave a white light, argon a bluish light, helium a yellow light, and neon, of course, a crimson light. Neon tubes for advertising became common in London by the 1920s, but it was only in the late 1930s that fluorescent tubes (using a mixture of mercury vapor with an inert gas) started to become a commercial possibility, a development in which Abe played a considerable part.

Uncle Dave, to show he was not bigoted, had a fluorescent light installed in his factory, and the two brothers, who had seen the tussle of gas and electricity in their youth, would sometimes argue about the respective merits and drawbacks of incandescent and fluorescent bulbs. Abe would say that filament bulbs would go the way of the gas mantle, Dave that fluorescents would always be bulky, never a match for the ease and cheapness of bulbs. (Both would have been surprised to find, fifty years later, that while fluorescents had evolved in all sorts of ways, filament bulbs remained as popular as ever, and that they coexisted in a comfortable and fraternal relationship.)

The more Uncle Abe showed me, the more mysterious the whole thing became. I understood a certain amount about light: that colors were how we saw different frequencies or wavelengths; and that the color of objects came from the way they absorbed or transmitted light, obstructing some frequencies, letting others through. I understood that black substances absorbed all the light, letting nothing through; and that with metals and mirrors it was the opposite — the wave front of light particles, as I imagined it, hit the mirror like a rubber ball and was reflected in a sort of instant bounce.

But none of these notions was helpful when one came to the phenomena of fluorescence and phosphorescence, for here one could shine an invisible light, a "black" light, on something and it would glow white or red or green or yellow, emitting a light of its own, a frequency of light not present in the illuminant.

And then there was the question of delay. The action of light normally seemed instantaneous. But with phosphorescence, the energy of sunlight, seemingly, was captured, stored, transformed into energy of a different frequency, and then emitted in a slow dribble, over hours (there were similar delays, Uncle Abe told me, with fluorescence, though these were far shorter, just fractions of a second). How was this possible?

19

MA

One summer after the war, in Bournemouth, I managed to obtain a very large octopus from a fisherman and kept it in the bath in our hotel room, which I filled with seawater. I would feed it live crabs, which it tore open with its horny beak, and I think it grew quite attached to me. It certainly recognized me when I came into the bathroom, and would flush different colors, indicating its emotion. Although we had had dogs and cats at home, I had never had an animal of my own. Now I had, and I thought my octopus quite as intelligent, and as affectionate, as any dog. I wanted to bring it back to London, give it a home, a huge tank festooned with sea anemones and seaweed, have it as my very own pet.

I did a lot of reading about aquariums and artificial seawater—but, in the event, the decision was taken from me, for one day the maid came in, and seeing the octopus in the bath, she had hysterics and poked it, wildly, with a long broom. The octopus, upset, discharged a huge cloud of ink, and when I returned a little later, I found it dead, sprawled out in its own ink. I dissected it, sorrowfully, when I got back to London, to learn what I could, and kept its scattered remains in formalin in my bedroom for many years.

Living in a medical household, hearing my parents and older brothers talk about patients and medical conditions, both fascinated and (sometimes) appalled me, but my new chemical vocabulary allowed me, in a sense, to compete with them. They might talk about *empyema* (a beautiful, nuggety, four-syllable word for a vile suppuration in the chest cavity), but I could cap it with *empyreuma,* that glorious word for the smell of burning organic matter. It was not just the sound of these words that I loved, but their etymology—I was now doing Greek and Latin at school, and I spent hours teasing out the origins and derivations of chemical terms, the sometimes twisted and indirect paths by which they had acquired their present meanings.

Both my parents were given to telling medical stories— stories which might start from a description of a pathological condition or an operation, and extend from this to an entire biography. My mother, especially, would tell such stories, to her students and colleagues, to dinner guests, or to anyone who was around; the medical, for her, was always embedded in a life. I would occasionally see the milkman or the gardener transfixed, listening to one of her clinical tales.

There was a large bookcase full of medical books in the surgery, and I would rummage through them at random, often in a state of mixed fascination and horror. Some of them I returned to again and again: there was Bland-Sutton's *Tumours Innocent and Malignant*—this was especially notable for its line drawings of monstrous teratomas and tumors; Siamese twins joined in the middle; Siamese twins with their faces fused together; two-headed calves; a baby with a tiny accessory head near its ear (a head which reflected, in tiny replica, I read, the expressions of the main face); "trichobezoars"—bizarre masses full of hair and other stuff, swallowed and embedded, sometimes fatally, in the stomach; an ovarian cyst so large it had to be carried on a handcart; and, of course, the Elephant Man, whom my father had already told me about (he had been a student at the London Hospital not so many years after John Merrick had lived there). Scarcely less horrifying was an *Atlas of Dermachromes,* showing every vile skin condition on the face of the earth. But the most informative, the most read, was French's *Differential Diagnosis*—its tiny line illustrations were especially appealing to me. Here, too, horrors lay in wait: the most frightening, for me, being the entry on progeria, a galloping senility that could hurtle a ten-year-old child through a lifetime within months, turning him into a fragile-boned, bald, beak-nosed, piping creature who looked as old as the shriveled, monkeylike Gagool—the three-hundred-year-old witch in *King Solomon's Mines*—or the demented Struldbrugs of Luggnagg.

Though with my return to London and my "apprenticeship" (as I sometimes imagined it) with my uncles, many of the fears of Braefield had vanished like a bad dream, they had left a residue of fear and superstition, a sense that some special awfulness might be reserved for me, and that this might descend at any moment.

The special dangers of chemistry were sought out, to some degree, I suspect, as a means of playing with such fears, persuading myself that by care and vigilance, prudence, forethought, one could learn to control, or find a way through, this hazardous world. And here, indeed, through care (and luck), I never hurt myself too much, and could maintain a sense of mastery and control. But with regard to life and health generally, no such protection could be counted on. Different forms of anxiety, of fearfulness, now struck me: I became afraid of horses (still used by the milkman to drive his float), afraid they might bite me with their large teeth; afraid of crossing the road, especially after our dog, Greta, was killed by a motorbike; afraid of other children, who (if nothing else) would laugh at me; afraid of stepping on the cracks between paving stones; and afraid, above all, of disease, of death.

My parents' medical books nourished these fears, fed an incipient disposition to hypochondria. Around the age of twelve I developed a mysterious, though hardly life-threatening, skin condition, which produced an exudation of serum behind my elbows and knees, stained my clothing, and made me avoid ever being seen undressed. Was it my fate, I wondered fearfully, to get one of these skin disorders or monstrous tumors I had read about—or was progeria the unspeakable fate that was reserved for me?

I was fond of the Morrison table, a huge table of iron that was lodged in the breakfast room, and which was strong enough, supposedly, to bear the whole weight of the house if we were bombed. There had been many accounts of such tables saving the lives of people who would otherwise have been crushed or suffocated beneath the debris of their own houses. The whole family would take cover under the table during air raids, and the notion of this protection, this shelter, took on an almost human character for me. The table would protect us, look out for us, care for us.

It was very cozy, I felt, almost a little cottage within the house, and when I came back from St. Lawrence College, at the age of ten, I would sometimes crawl under it and sit or lie there, quietly, even when there was not an air raid.

My parents recognized that I was in a fragile state at this time, and they offered no comments when I retreated and crawled under the table. But one evening, when I emerged from beneath it, they were horrified to see a bald circle on my scalp—ringworm was their instant, medical diagnosis. My mother scrutinized me more closely and murmured with my father. They had never heard of ringworm appearing with such suddenness. I admitted nothing, tried to look innocent, and concealed the razor, Marcus's razor, which I had taken with me under the table. The next day, they took me to a skin specialist, a Dr. Muende. Dr. Muende gave me a piercing glance—I had no doubt that he saw straight through me—took a specimen of hair from the bare spot, and put it under his microscope to examine. After a second, he said, *"Dermatitis artefacta,"* meaning that the hair loss was self-inflicted, and when he said this, I blushed a deep crimson. There was no discussion afterwards as to why I had shaved my head, or lied.

My mother was an intensely shy woman who could hardly bear social occasions and would retreat into silence, or her own thoughts, when forced into them. But there was another side to her character, and she could become expansive, exuberant, a ham, a performer, when she was at ease, with her students. Many years later, when I took my first book to an editor at Faber's, she said, "You know, we've met before."

"I don't think I remember," I said, embarrassed. "I can never recognize faces."

"You wouldn't," she rejoined. "It was many years ago, when I was a student of your mother's. She was lecturing on breast-feeding that day, and after a few minutes she suddenly broke off, saying, 'There's nothing too difficult or embarrassing about breastfeeding.' She bent down and retrieved a small baby which had been sleeping, concealed behind her desk, and, unwrapping the infant, breastfed it before the class. It was in September 1933, and you were the infant."

I, too, have my mother's shyness, her dread of social occasions, as well as her flamboyance, her exuberance in front of an audience, in equal measure.

There was another level, a deeper level, in her, a realm of total absorption in her work. Her concentration when she was operating was absolute (though she might break the almost religious silence at times by telling a joke or giving a recipe to one of her assistants). She had an intense feeling for structure, the way things were put together—whether they were human bodies, or plants, or scientific instruments or machines. She still had the microscope, an old Zeiss, she had had as a student, and kept it polished and oiled and in perfect shape. She still enjoyed sectioning specimens, hardening them, fixing them, staining them with different dyes—the whole intricate panoply of techniques used to make sectioned tissues stable and richly visible. She introduced me to some of the wonders of histology with these slides, and I came to recognize—in the brilliant stains of hemotoxylin and eosin, or blackly shadowed with osmium—a variety of cells both healthy and malignant. I could appreciate the abstract beauty of these slides without worrying too much about the disease or surgery that had brought them into being. I loved, too, the odorous gums and liquids used to make them; the smells of clove oil, cedarwood oil, Canada balsam, xylene are still associated, in my mind, with the memory of my mother, intently bending over her microscope, totally absorbed.

Though both my parents were highly sensitive to the sufferings of their patients—more so, I sometimes thought, than to those of their children—their orientations, their perspectives, were fundamentally different. My father's quiet hours were all spent with books, in the library, surrounded by biblical commentaries or occasionally his favorite First World War poets. Human beings, human behavior, human myths and societies, human language and religions occupied his entire attention—he had little interest in the nonhuman, in "nature," as my mother had. I think my father was drawn to medicine because its practice was central in human society, and that he saw himself in an essentially social and ritual role. I think my mother, though, was drawn to medicine because for her it was part of natural history and biology. She could not look at human anatomy or physiology without thinking of parallels and precursors in other primates, other vertebrates. This did not compromise her concern and feeling for the individual—but placed it, always, in a wider context, that of biology and science in general.

Her love of structure extended in all directions. Our old grandfather clock, with its intricate dials and inner machinery, was very delicate, needing constant care. My mother undertook this entirely herself, becoming a sort of horologist in the process. Similarly with other things in the house, even the plumbing. There was nothing she liked more than mending a leaky faucet or a toilet, and the services of outside plumbers were usually not needed.

But her best hours, her happiest hours, were spent in the garden, and here her sense of structure and function, her esthetic feeling, and her tenderness came together—plants, after all, were living creatures, much more wonderful, but also more needy, than clocks or cisterns. When, years later, I came across the phrase "a feeling for the organism"—often used by the geneticist Barbara McClintock—I realized that this defined my mother exactly, and that this feeling for the organism underlay everything from her green thumb in the garden to the delicacy and success of her operations.

My mother loved the garden, the large plane trees that edged Exeter Road, the lilacs that filled it with their scent in May, and the climbing roses that trellised its brick walls. She gardened whenever she could, and was especially attached to the fruit trees she had planted—a quince tree, a pear tree, two crab apples, and a walnut. She was particularly fond of ferns as well, and the "flower" beds were almost entirely filled with them.

The conservatory, at the far end of the drawing room, was one of my favorite places, the place where, before the war, my mother had kept her most tender plants. It somehow escaped being shattered during the war, and when my own botanical interests blossomed, I shared it with her. I have tender memories of a tree fern, a woolly *Cibotium,* I tried to grow there in 1946, and a cycad, a *Zamia,* with stiff cardboard leaves.

Once, when my nephew Jonathan was a few months old, I picked up a packet of X-rays marked "J. Sacks" that had been left in the lounge. I started to leaf through them curiously, then perplexedly, then with horror—for Jonathan was a nice-looking little baby, and no one would have guessed, without the X-rays, that he was hideously deformed. His pelvis, his little legs—they scarcely looked human.

I went to my mother with the X-rays, shaking my head. "Poor Jonathan . . . " I started.

My mother looked puzzled. "Jonathan?" she said. "Jonathan is fine."

"But the X-rays," I said, "I've been looking at his X-rays."

My mother looked blank, then burst into a roar of laughter, and laughed until the tears ran down her face. "J" did not stand for Jonathan, she finally said, but for another member of the household, Jezebel. Jezebel, our new boxer, had had some blood in her urine, and my mother had taken her to hospital to have a kidney X-ray. What I had taken for grotesquely deformed human anatomy was, in fact, perfectly normal canine anatomy. How could I have made such an absurd mistake? The least knowledge, the least common sense, would have made it all clear to me—my mother, a professor of anatomy, shook her head in disbelief.

My mother's practice had moved, sometime in the 1930s, from general surgery to gynecology and obstetrics. There was nothing she loved more than a challenging delivery—an arm presentation, a breech—brought off successfully. But she would occasionally bring back malformed fetuses to the house—anencephalic ones with protruding eyes at the top of their brainless, flattened heads, or spina bifida ones in which the whole spinal cord and brainstem were exposed. Some of these had been still-born, others she and the matron had quietly drowned at birth ("like a kitten," she once said), feeling that if they lived, no conscious or mental life would ever be possible for them. Eager that I should learn about anatomy and medicine, she dissected several of these for me, and then insisted, though I was only eleven, that I dissect them myself. She never perceived, I think, how distressed I became, and probably imagined that I was as enthusiastic here as she was. Though I had taken to dissecting naturally, by myself, with earthworms and frogs and with my octopus, the dissection of these human fetuses filled me with revulsion. My mother often told me how she had worried about the growth of my own skull as an infant, fearing that the fontanelles had closed too early, and that I would be, in consequence, a microcephalic idiot. Thus I saw in these fetuses what (in imagination) I, too, could have been, and this made it more difficult to distance myself, and heightened my horror.

Though it was understood, almost from my birth, that I would be a doctor (and specifically, my mother hoped, a surgeon), these precocious experiences turned me against medicine, made me want to escape and turn to plants, which had no feelings, to crystals and minerals and elements, above all, for they existed in a deathless realm of their own, where sickness and suffering, pathology, held no sway.

When I was fourteen, my mother arranged with a colleague, a professor of anatomy at the Royal Free Hospital, that I should be inducted into human anatomy, and Professor G., agreeable, took me to the dissecting room. There, on long trestles, lay the bodies, wrapped in yellow oilskin (to keep the exposed tissues from drying when they were not being dissected). This was the first time I had ever seen a corpse, and the bodies looked oddly shrunken and small to me. There was a horrible smell of mortified tissue and preservative in the air, and I came near to fainting as I went in—there were spots before my eyes and a sudden surge of nausea. Professor G. said she had selected a body for me, the body of a fourteen-year-old girl. Some of the girl had been dissected already, but there was a nice, untouched leg I could start on. I had a desire to ask who the girl was, what she had died of, what had brought her to such a pass—but Professor G. proferred no information, and in a way I was glad, for I dreaded to know. I had to think of this as a cadaver, a nameless thing of nerves and muscles, tissues and organs, to be dissected as one would dissect a worm or a frog, to learn how the organic machine was put together. There was a manual of anatomy, Cunningham's *Manual,* at the head of the table; this was the copy the medical students used when they were dissecting, and its pages were yellowed and greasy with human fat.

My mother had bought me a Cunningham the week before, so that I had some knowledge, but this in no way prepared me for the actual experience, the emotional experience, of dissecting my first body. Professor G. started me off with a broad initial incision down the thigh, parting the fat and exposing the fascia beneath. She gave me various tips, then she thrust the scalpel into my hand—she would be back in half an hour, she said, to see how I was getting on.

It took me a month to dissect the leg; the most difficult was the foot, with its little muscles and stringy tendons, and the knee joint, with all its complexity. Occasionally I could feel how beautifully everything was put together, could enjoy an intellectual and esthetic pleasure such as my mother got from surgery and anatomy. Her own professor, in her medical student days, had been the famous comparative anatomist Frederic Wood-Jones. She loved the books he had written—*Arboreal Man, The Hand,* and *The Foot*—and cherished the copies he had inscribed for her. She was amazed when I said I could not "understand" the foot. "But it is like an arch," she said, and started drawing feet— drawings an engineer might have done, from every angle, to show how the foot combined stability with flexibility, how beautifully designed or evolved it was for walking (though bearing obvious residues of its original, prehensile function too).

I lacked my mother's powers of visualization, her strong mechanical and engineering sense, but I loved it when she talked of the foot and drew, in rapid succession, the feet of lizards and birds, horses' hooves, lions' paws, and a series of primate feet. But this delight in understanding and appreciating anatomy was lost, for the most part, in the horror of the dissection, and the feeling of the dissecting room spread to life outside—I did not know if I would ever be able to love the warm, quick bodies of the living after facing, smelling, cutting the formalin-reeking corpse of a girl my own age.

20

PENETRATING RAYS

It was in Abe's attic that I was introduced to cathode rays. He had a highly efficient vacuum pump, and an induction coil — a two-foot-long cylinder wound with miles and miles of densely coiled copper wire and set on a mahogany base. There were two large moveable brass electrodes above the coil, and when the coil was switched on, there would leap between them a formidable spark, a miniature lightning bolt, something out of Dr. Frankenstein's lab. Uncle showed me how he could separate the electrodes until they were too far apart to spark and then connect them to a yard-long vacuum tube. As he reduced the pressure in the electrified tube, a series of extraordinary phenomena appeared inside it: first a flickering light with red streamers like a miniature Aurora Borealis, then a brilliant column of light filling the whole tube. As the pressure was lowered still further, the column broke into discs of light separated by dark spaces. Finally, at a ten-thousandth of an atmosphere, everything became dark again inside the tube, but the end of the tube itself started to glow with a brilliant fluorescence. The tube was now filled, Uncle said, with cathode rays, little particles shot off the cathode with a tenth the speed of light, and so energetic that if one converged them with a saucer-shaped cathode they could heat a piece of platinum foil to red heat. I was a little afraid of these cathode rays (as I had been, as a child, of the ultraviolet rays in the surgery), for they were both potent and invisible, and I wondered if they could leak out of the tube and dart at us, unseen, in the darkened attic.

Cathode rays, Uncle Abe assured me, could travel only two or three inches in ordinary air—but there was another sort of ray, far more penetrating, which Wilhelm Roentgen had discovered in 1895 while experimenting with just such a cathode-ray tube. Roentgen had covered the tube with a cylinder of black cardboard to prevent any leakage of cathode rays, and yet he was astounded to observe that a screen painted with a fluorescent substance lit up brilliantly with each discharge of the tube, even though it was halfway across the room.

Roentgen immediately decided to drop his other research projects in order to investigate this totally unexpected and almost incredible phenomenon, repeating the experiment over and over again to convince himself that the effect was authentic. (He told his wife that if he spoke of it without the most convincing evidence, people would say, "Roentgen has gone mad.") For the next six weeks he investigated the properties of these extraordinarily penetrating new rays and found that, unlike visible light, they could not, apparently, be refracted or diffracted. He tested their ability to pass through all sorts of solids and found they could pass to some degree through most common materials and still activate a fluorescent screen. When Roentgen placed his own hand in front of the fluorescing screen, he was astonished to see a ghostly silhouette of its bones. A set of metal weights, similarly, became visible through their wooden box—wood and flesh were more transparent to the rays than metal or bone. The rays also affected photographic plates, he found, so that in his first paper he was able to publish photographs taken by the X-rays, as he called them—including a radiograph of his wife's hand, her wedding ring encircling a skeletal finger.

On January 1, 1896, Roentgen published his findings and first radiographs in a small academic journal. Within days the major newspapers of the world picked up the story. The sensational impact of his discovery horrified the shy Roentgen, and after his initial paper and a verbal presentation the same month, he never discussed X-rays again, but returned to working quietly on the varied scientific interests which had engrossed him in the years before 1896. (Even when he was awarded the first Nobel Prize in physics in 1901 for his discovery of X-rays, he declined to give a Nobel speech.)

But the usefulness of this new technology was readily apparent, and X-ray facilities were rapidly set up around the world for medical use—to detect fractures, find foreign bodies, gallstones, etc. By the end of 1896, more than a thousand scientific articles on X-rays had appeared. The response to Roentgen's rays, indeed, was not only medical and scientific, but seized the public imagination in various ways. One could buy, for a dollar or two, an X-ray photograph of a nine-week-old infant "showing with beautiful detail the bones of the skeleton, the stage of ossification, the location of the liver, stomach, heart, etc."

X-rays, it was felt, might have the power to penetrate the most intimate, hidden, secret parts of people's lives. Schizophrenics felt that their minds could be read or influenced by X-rays; others felt that nothing was safe. "You can see other people's bones with the naked eye," thundered one editorial, "and also through eight inches of solid wood. On the revolting indecency of this there is no need to dwell." Lead-lined underclothes were put on sale to shield people's private parts from the all-seeing rays. A ditty appeared in the journal *Photography*, ending,

> I hear they'll gaze
> through cloak and gown—and even stays,
> those naughty, naughty, Roentgen rays.

My uncle Yitzchak, after being in practice with my father during the months of the great flu epidemic, had been drawn into the practice of radiology soon after the First World War. He had gone on, my father told me, to gain uncanny powers of diagnosis by X-ray, able almost unconsciously to pick up the smallest hints of any pathological process.

In his consulting rooms, which I visited a few times, Uncle Yitzchak showed me something of his apparatus and its uses. The X-ray tube in his machine was no longer visible, as it had been in the early machines, but was housed in a beaked and humped black metal box—it looked rather dangerous and predatory, like the head of a giant bird. Uncle Yitzchak took me into the darkroom to watch him develop an X-ray he had just taken. Dimly, in the red light, translucent almost, beautiful, I saw the outlines of a thighbone, a femur, on the large film. Uncle pointed out to me a tiny hairline fracture, just visible as a grey line.

"You've seen X-ray screening," said Uncle Yitzchak, "in shoe shops, which show you the bones moving through the flesh.[1] We can also use special contrast media to show us some of the other tissues in the body—it's marvelous!"

---

[1] Shoe shops everywhere in my boyhood were equipped with X-ray machines, fluoroscopes, so that one could see how the bones of one's feet were fitting in new shoes. I loved these machines, for one could wiggle one's toes and see the many separate bones in the foot moving in unison, in their almost transparent envelope of flesh.

Uncle Yitzchak asked if I would like to watch this. "You remember Mr. Spiegelman, the mechanic? Your father suspects that he has a stomach ulcer, and sent him to me to find out. He's going to have a barium 'meal.'

"We use barium sulfate," Uncle continued, stirring up the heavy white paste, "because barium ions are heavy and almost opaque to X-rays." This comment intrigued me and made me wonder why one could not use even heavier ions instead. One could have, perhaps, a lead, or mercury, or thallium "meal"—all of these had exceptionally heavy ions, though, of course, the meals would be lethal. A gold or platinum meal would be fun, but far too expensive. "What about a tungsten meal?" I suggested. "Tungsten atoms are heavier than barium, and tungsten is neither toxic nor expensive."

We entered the examining room, and Uncle introduced me to Mr. Spiegelman—he remembered me from one of our Sunday morning rounds. "This is Dr. Sacks's youngest, Oliver—he wants to be a scientist!" Uncle positioned Mr. Spiegelman between the X-ray machine and a fluorescent screen and gave him the barium meal to eat. Mr. Spiegelman spooned the paste down, grimacing, and started to swallow it, as we watched on the screen. As the barium passed down the throat and into the esophagus, I could see this filling and writhing, slowly, as it pushed the bolus of barium into the stomach. I could see, more faintly, a ghostlike background, the lungs expanding and contracting with each breath. Most disconcerting of all, I could see a sort of bag, pulsing—that, Uncle said, pointing, was the heart.

I had sometimes wondered what it would be like to have other senses. My mother had told me that bats used ultrasound, that insects saw ultraviolet, that rattlesnakes could sense infrared. But now, watching Mr. Spiegelman's innards exposed to the X-ray "eye," I was glad that I did not have X-ray vision myself, and that I was confined, by nature, to a small part of the spectrum.

Like Uncle Dave, Uncle Yitzchak retained a strong interest in the theoretical foundations of his subject and its historical development, and he also had a little "museum," in this case of old X-ray and cathode-ray tubes, going back to the fragile, three-pronged ones that had been used in the 1890s. The early tubes, Yitzchak said, offered no protection against stray radiation, nor were the dangers of radiation fully realized in the early days. And yet, he added, X-rays had shown their dangers from the start: skin burns were seen within months of their introduction, and Lord Lister himself, the discoverer of antisepsis, issued a warning as early as 1896—but it was a warning that no one heeded.[2]

It was also apparent from the start that X-rays carried a good deal of energy and would generate heat wherever they were absorbed. Yet, penetrating as they were, X-rays did not have too great a range in air. It was the opposite with wireless waves, radio waves, which, if properly projected, could leap across the Channel with the speed of light. These, too, carried energy. I wondered whether these strange, sometimes dangerous relatives of visible light had perhaps suggested to H. G. Wells the sinister heat ray used by the Martians in *The War of the Worlds,* published only two years after Roentgen's discovery. The Martian heat ray, Wells wrote, was "the ghost of a beam of light," "an invisible yet intensely heated finger," "an invisible, inevitable sword of heat." Projected by a parabolic mirror, it would soften iron, melt glass, make lead run like water, make water explode incontinently into steam. And its passage across the countryside, Wells added, was "as swift as the passage of light."

---

[2] Dentists were especially at risk, holding small X-ray films inside their patients' mouths, often for minutes at a time, for the original emulsions were very slow. Many dentists lost fingers by exposing their hands to X-rays in this way.

. . .

While X-rays took off, engendering innumerable practical applications and perhaps an equal number of fantasies, they elicited a very different train of thought in the mind of Henri Becquerel. Becquerel was already distinguished in many fields of optical research, and came from a family in which a passionate interest in luminescence had been central for sixty years.[3] He was intrigued when he heard in early 1896 the first news of Roentgen's X-rays and the fact that they seemed to be emanating not from the cathode itself but from the fluorescent spot where the cathode rays hit the end of the vacuum tube. He wondered whether the invisible X-rays might not be a special form of energy that went along with the visible phosphorescence—and whether indeed all phosphorescence might be accompanied by the emission of X-rays.

Since no substances fluoresced more brilliantly than uranium salts, Becquerel pulled out a specimen of a uranium salt, potassium uranyl sulfate, exposed it to the sun for several hours, and then laid it on a photographic plate wrapped in black paper. He was greatly excited to find that the plate was darkened by the uranium salt, even through the paper, just as with X-rays, and that a "radiograph" of a coin could be easily obtained.

Becquerel wanted to repeat his experiment, but (this was the middle of the Parisian winter and the sky remained overcast) he was unable to expose the uranium salt to the sun, so it lay undisturbed in the drawer for a week, on top of the black-wrapped

---

[3] Henri Becquerel's grandfather, Antoine Edmond Becquerel, had launched the systematic study of phosphorescence in the 1830s and published the first pictures of phosphorescent spectra. Antoine's son, Alexandre-Edmond, had assisted in his father's research and invented a "phosphoroscope," which allowed him to measure fluorescences that lasted as briefly as a thousandth of a second. His 1867 book, *Lumière,* was the first comprehensive treatment of phosphorescence and fluorescence to appear (and the only one for the next fifty years).

photographic plate, with a small copper cross in between. But then, for some reason—was it an accident, or a premonition?—he developed the photographic plate anyway. It was darkened as strongly as if the uranium had been exposed to sunlight, indeed more so, and showed a clear silhouette of the copper cross.

Becquerel had discovered a new and much more mysterious power than Roentgen's rays—the power of a uranic salt to emit a penetrating radiation that could fog a photographic plate, and in a way that had nothing to do with exposure to light or X-rays or, seemingly, any other external source of energy. Becquerel, his son later wrote, was "stupefied" at this finding ("*Henri Becquerel fut stupefait*")—as Roentgen had been by his X-rays—but then, like Roentgen, he investigated the "impossible." He found that the rays retained all their potency even if the uranic salt was kept for two months in a drawer; and that they had the power not only to darken photographic plates but also to ionize air, render it conducting, so that electrically charged bodies in their vicinity would lose their charge. This indeed provided a very sensitive way of measuring the intensity of Becquerel's rays, using an electroscope.

Investigating other substances, he found that this power was possessed not only by uranic salts but uranous ones too, even though these were not phosphorescent or fluorescent. On the other hand, barium sulfide, zinc sulfide, and certain other fluorescent or phosphorescent substances had no such power. Thus the "uranium rays," as Becquerel now called them, had nothing to do with fluorescence or phosphorescence as such—and everything to do with the element uranium. They had, like X-rays, a very considerable power of penetrating materials opaque to light, but unlike X-rays, they were apparently emitted spontaneously. What were they? And how could uranium continue to radiate them, with no apparent diminution, for months at a time?

Uncle Abe encouraged me to repeat Becquerel's discovery in my own lab, giving me a chunk of pitchblende rich in uranium oxide. I took the heavy chunk home, wrapped in lead foil, in my school satchel. The pitchblende had been sectioned cleanly through the middle, to show its structure, and I placed the cut face flat on some film—I had begged a sheet of special X-ray film from Uncle Yitzchak, and I kept this wrapped in its dark paper. I left the pitchblende lying on the covered film for three days, then took it along to him to develop. I was wild with excitement when Uncle Yitzchak developed it in front of me, for now one could see the glares of radioactivity in the mineral—radiation and energy whose existence, without the film, one would never have guessed at.

I was doubly thrilled by this, because photography was becoming a hobby, and I now had my first picture taken by invisible rays! I had read that thorium, too, was radioactive, and, knowing that gas mantles contained this, I detached one of the delicate, thoria-impregnated mantles at home from its base and carefully spread it over another piece of X-ray film. This time I had to wait longer, but after two weeks I got a beautiful "autoradiograph," the fine texture of the mantle picked out by the thorium rays.

Though uranium had been known since the 1780s, it had taken more than a century before its radioactivity was discovered. Radioactivity might have been discovered, perhaps, in the eighteenth century, had anyone chanced to place a piece of pitchblende close to a charged Leyden jar or an electroscope. Or it might have been discovered in the middle of the nineteenth century, had a piece of pitchblende, or some other uranium ore or salt, been left in accidental proximity to a photographic plate. (This had in fact happened to one chemist, who, not realizing what had happened, sent the plates back to the manufacturer with an indignant note saying that they were "spoilt.") Yet had radioactivity been discovered earlier, it would have been seen simply as a curiosity, a freak, a *lusus naturae,* its enormous significance wholly unsuspected. Its discovery would have been premature, in the sense that there would have been no nexus of knowledge, no context, to give it meaning. Indeed, when radioactivity was finally discovered in 1896, there was very little reaction at first, for even then its significance could barely be grasped. So in contrast to Roentgen's discovery of X-rays, which instantly captured the public's attention, Becquerel's discovery of uranium rays was virtually ignored.

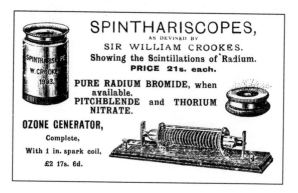

MADAME CURIE'S ELEMENT

My mother worked at many hospitals, including the Marie Curie Hospital in Hampstead, a hospital that specialized in radium treatments and radiotherapy. I was not too sure, as a child, what radium was, but I understood it had healing powers and could be used to treat different conditions. My mother said the hospital possessed a radium "bomb." I had seen pictures of bombs and read about them in my children's encyclopedia, and I imagined this radium bomb as a great winged thing that might explode at any moment. Less alarming were the radon "seeds" which were implanted in patients—little gold needles full of a mysterious gas—and once or twice she brought an exhausted one home. I knew my mother admired Marie Curie hugely—she had met her once, and would tell me, even when I was quite small, how the Curies had discovered radium, and how difficult this had been, because they had had to work through tons and tons of heavy mineral ore to get the merest speck of it.

Eve Curie's biography of her mother—which my own mother gave me when I was ten—was the first portrait of a scientist I ever read, and one that deeply impressed me.[1] It was no dry recital of a life's achievements, but full of evocative, poignant images—Marie Curie plunging her hands into the sacks of pitchblende residue, still mixed with pine needles from the Joachimsthal mine; inhaling acid fumes as she stood amid vast steaming vats and crucibles, stirring them with an iron rod almost as big as herself; transforming the huge, tarry masses to tall vessels of colorless solutions, more and more radioactive, and steadily concentrating these, in turn, in her drafty shed, with dust and grit continually getting into the solutions and undoing the endless work. (These images were reinforced by the film *Madame Curie,* which I saw soon after reading the book.)

Even though the rest of the scientific community had ignored the news of Becquerel's rays, the Curies were galvanized by it: this was a phenomenon without precedent or parallel, the revelation of a new, mysterious source of energy; and nobody, apparently, was paying any attention to it. They wondered at once whether there were any substances besides uranium that emitted similar rays, and started on a systematic search (not confined, as Becquerel's had been, to fluorescent substances) of everything they could lay their hands on, including samples of almost all the seventy known elements in some form or other. They found only one other substance besides uranium that emitted Becquerel's rays, another element of very high atomic weight—thorium.

---

[1] In 1998 I spoke at a meeting for the centennial of the discovery of polonium and radium. I said that I had been given this book when I was ten, and that it was my favorite biography. As I was talking I became conscious of a very old lady in the audience, with high Slavic cheekbones and a smile going from one ear to the other. I thought, "It can't be!" But it was—it was Eve Curie, and she signed her book for me sixty years after it was published, fifty-five years after I got it.

Testing a variety of pure uranium and thorium salts, they found the intensity of the radioactivity seemed to be related only to the amount of uranium or thorium present; thus one gram of metallic uranium or thorium was more radioactive than one gram of any of their compounds.

But when they extended their survey to some of the common minerals containing uranium and thorium, they found a curious anomaly, for some of these were actually more active than the element itself. Samples of pitchblende, for instance, might be up to four times as radioactive as pure uranium. Could this mean, they wondered, in an inspired leap, that another, as-yet-unknown element was also present in small amounts, one that was far more radioactive than uranium itself?

In 1897 the Curies launched upon an elaborate chemical analysis of pitchblende, separating the many elements it contained into analytic groups: salts of alkali metals, of alkaline earth elements, of rare-earth elements — groups basically similar to those of the periodic table — to see if the unknown radioactive element had chemical affinities with any of them. Soon it became clear that a good part of the radioactivity could be concentrated by precipitation with bismuth.

They continued rendering their pitchblende residue down, and in July of 1898 they were able to make a bismuth extract four hundred times more radioactive than uranium itself. Knowing that spectroscopy could be thousands of times more sensitive than traditional chemical analysis, they now approached the eminent rare-earth spectroscopist Eugène Demarçay to see if they could get a spectroscopic confirmation of their new element. Disappointingly, no new spectral signature could be obtained at this point; but nonetheless, the Curies wrote,

> we believe the substance we have extracted from pitchblende contains a metal not yet observed, related to bismuth by its analytical properties. If the existence of this new metal is confirmed

we propose to call it polonium, from the name of the original country of one of us.

They were convinced, moreover, that there must be still another radioactive element waiting to be discovered, for the bismuth extraction of polonium accounted for only a portion of the pitchblende's radioactivity.

They were unhurried—no one else, after all, it seemed, was even interested in the phenomenon of radioactivity, apart from their good friend Becquerel—and at this point took off on a leisurely summer holiday. (They were unaware at the time that there was another eager and intense observer of Becquerel's rays, the brilliant young New Zealander Ernest Rutherford, who had come to work in J. J. Thomson's lab in Cambridge.) In September the Curies returned to the chase, concentrating on precipitation with barium—this seemed particularly effective in mopping up the remaining radioactivity, presumably because it had close chemical affinities with the second as-yet-unknown element they were now seeking. Things moved swiftly, and within six weeks they had a bismuth-free (and presumably polonium-free) barium chloride solution which was nearly a thousand times as radioactive as uranium. Demarçay's help was sought once again, and this time, to their joy, he found a spectral line (and later several lines: "two beautiful red bands, one line in the blue-green, and two faint lines in the violet") belonging to no known element. Emboldened by this, the Curies claimed a second new element a few days before the close of 1898. They decided to call it radium, and since there was only a trace of it mixed in with the barium, they felt its radioactivity "must therefore be enormous."

It was easy to claim a new element: there had been more than two hundred such claims in the course of the nineteenth century, most of which turned out to be cases of mistaken identity, either "discoveries" of already known elements or mixtures of elements. Now, in a single year, the Curies had claimed the existence of not one but two new elements, solely on the basis of a heightened radioactivity and its material association with bismuth and barium (and, in the case of radium, a single new spectral line). Yet neither of their new elements had been isolated, even in microscopic amounts.

Pierre Curie was fundamentally a physicist and theorist (though dexterous and ingenious in the lab, often devising new and original apparatus—one such was an electrometer, another a delicate balance based on a new piezo-electric principle—both subsequently used in their radioactivity studies). For him, the incredible phenomenon of radioactivity was enough—it invited a vast new realm of research, a new continent where countless new ideas could be tested.

But for Marie, the emphasis was different: she was clearly enchanted by the physicality of radium as well as its strange new powers; she wanted to see it, to feel it, to put it in chemical combination, to find its atomic weight and its position in the periodic table.

Up to this point the Curies' work had been essentially chemical, removing calcium, lead, silicon, aluminum, iron, and a dozen rare-earth elements—all the elements other than barium—from the pitchblende. Finally, after a year of this, there came a time when chemical methods alone no longer sufficed. There seemed no chemical way of separating radium from barium, so Marie Curie now began to look for a physical difference between their compounds. It seemed probable that radium would be an alkaline earth element like barium and might therefore follow the trends of the group. Calcium chloride is highly soluble; strontium chloride less so; barium chloride still less so—radium chloride, Marie Curie predicted, would be virtually insoluble. Perhaps one could make use of this to separate the chlorides of barium and radium, using the technique of fractional crystallization. As a warm solution is cooled, the less soluble solute will crystallize out first, and this was a technique which had been pioneered by the rare-earth chemists, striving to separate elements that were chemically almost indistinguishable. It was one that required great patience, for hundreds, even thousands, of fractional crystallizations might be needed, and it was this repetitive and tantalizingly slow process that now caused the months to extend into years.

The Curies had hoped they might isolate radium by 1900, but it was to take nearly four years from the time they announced its probable existence to obtain a pure radium salt, a decigram of radium chloride—less than a ten-millionth part of the original. Fighting against all manner of physical difficulties, fighting the doubts and skepticisms of most of their peers, and sometimes their own hopelessness and exhaustion; fighting (although they did not know it) against the insidious effects of radioactivity on their own bodies, the Curies finally triumphed and obtained a few grains of pure white crystalline radium chloride—enough to calculate radium's atomic weight (226), and to give it its rightful place, below barium, in the periodic table.

To obtain a decigram of an element from several tons of ore was an achievement with no precedent; never had an element been so hard to obtain. Chemistry alone could not have succeeded in this, nor could spectroscopy alone, for the ore had to be concentrated a thousandfold before the first faint spectral lines of radium could even be seen. It had required a wholly new approach—the use of radioactivity itself—to identify the infinitesimal concentration of radium in its vast mass of surrounding material, and to monitor it as it was slowly, reluctantly, forced into a state of purity.

With this achievement, public interest in the Curies exploded, spreading equally to their magical new element and the romantic, heroic husband-and-wife team who had dedicated themselves so totally to its exploration. In 1903, Marie Curie summarized the work of the previous six years in her doctoral thesis, and in the same year she received (with Pierre Curie and Becquerel) the Nobel Prize in physics.

Her thesis was immediately translated into English and published (by William Crookes in his *Chemical News*), and my mother had a copy of this in the form of a little booklet. I loved the minute descriptions of the elaborate chemical processes the Curies performed, the careful, systematic examination of radium's properties, and especially the sense of intellectual excitement and wonder that seemed to simmer beneath the even-toned scientific prose. It was all down-to-earth, even prosaic—but it was a sort of poetry, too. And I was attracted by the notices on its covers for radium, thorium, polonium, uranium—all of these were freely available, to anyone, for fun or experiment.

There was an advertisement from A. C. Cossor, in Farringdon Road, a few doors from Uncle Tungsten's place, selling "pure radium bromide (when available), pitchblende . . . Crooke's high-vacuum tubes, showing the fluorescence of various minerals . . . [and] other scientific materials." Harrington Brothers (in Oliver's Yard, not far away) sold a variety of radium salts and uranium minerals. J. J. Griffin and Sons (later to become Griffin & Tatlock, where I went for my own chemical supplies) were selling "Kunzite—the new mineral, responding in a high degree to the emanations from radium," while Armbrecht, Nelson & Co. (a cut above the rest, in Grosvenor Square) had polonium sulfide (in tubes of one gram, twenty-one shillings) and screens of fluorescent willemite (sixpence for a square inch). "Our newly invented Thorium inhalers," they added, "may be had on hire." What, I wondered, was a thorium inhaler? Would one feel braced, strengthened, inhaling the radioactive element?

No one seemed to have any idea of the danger of these stuffs at this time.[2] Marie Curie herself mentioned in her thesis how "if a radio-active substance is placed in the dark in the vicinity of the

---

[2] Becquerel had been the first to note the injury that might result from radioactivity—he discovered a burn on himself after carrying a highly radioactive concentrate in his waistcoat pocket. Pierre Curie explored the matter, allowing a deliberate radium burn on his arm. Yet he and Marie never fully

closed eye or of the temple, a sensation of light fills the eye," and I often tried this myself, using one of the luminous clocks in our house, their figures and hands painted with Uncle Abe's luminous paint.

I was particularly moved by the description in Eve Curie's book of how her parents, restless one evening and curious as to how the fractional crystallizations were going, returned to their shed late one night and saw in the darkness a magical glowing everywhere, from all the tubes and vessels and basins containing the radium concentrates, and realized for the first time that their element was spontaneously luminous. The luminosity of phosphorus required the presence of oxygen, but the luminosity of radium arose entirely from within, from its own radioactivity. Marie Curie wrote in lyrical terms of this luminosity:

> One of our joys was to go into our workroom at night when we
> perceived the feebly luminous silhouettes of the bottles and cap-

---

faced the dangers of radium, their "child." Their laboratory, it was said, glowed in the dark, and both, perhaps, were to die from its effects. (Pierre, weakened, died in a traffic accident; Marie, thirty years later, from an aplastic anemia.) Radioactive specimens were sent freely in the post, and handled with little protection. Frederick Soddy, who worked with Rutherford, believed that handling radioactive materials had made him sterile.

And yet there was ambivalence, for radioactivity was also seen as benign, as healing. Besides thorium inhalers, there was thorium toothpaste, made by the Auer Company (Auntie Annie used to keep her dentures overnight in a glass with "radium sticks"), and the Radioendocrinator, containing radium and thorium, to be worn around the neck to stimulate the thyroid or around the scrotum to stimulate the libido. People went to spas to take the radium water.

The most serious problem arose in the United States, where doctors prescribed the drinking of radioactive solutions such as Radithor as rejuvenating agents, as well as to cure stomach cancer or mental illness. Thousands of people drank such potions, and it was only the highly publicized death in 1932 of Eben Byers, a prominent steel magnate and socialite, that put an end to the radium craze. After consuming a daily radium tonic for four years, Byers developed severe radiation sickness and cancer of the jaw; and he died grotesquely as his bones disintegrated, like Monsieur Valdemar in the Edgar Allan Poe story.

sules containing our products. . . . It was really a lovely sight and always new to us. The glowing tubes looked like faint fairy lights.

Uncle Abe still had some radium in his possession, left over from his work on luminous paint, and he would show me this, pulling out a vial with a few milligrams of radium bromide—it appeared to be a grain of ordinary salt—at the bottom. He had three little screens painted with platinocyanides—lithium, sodium, and barium platinocyanide—and as he waved the tube of radium (gripped in a pair of tongs) near the darkened screens, these lit up suddenly, becoming sheets of red, then yellow, then green fire, each fading suddenly as he moved the tube away again.

"Radium has lots of interesting effects on substances around it," he said. "The photographic effects you know, but radium also browns paper, burns it, pits it, like a colander. Radium decomposes the atoms of the air, and then they recombine in different forms—so you smell ozone and nitrogen peroxide when you are around it. It affects glass—it turns soft glasses blue, and hard glasses brown; it can also color diamonds and turn rock salt a deep, intense violet." Uncle Abe showed me a piece of fluorspar which he had exposed to radium for a few days. Its original color had been purple, he said, but now it was pale, charged with strange energy. He heated the fluorspar a little, far below red heat, and it suddenly gave off a brilliant flash, as if it were white-hot, and returned to its original purple.

Another experiment Uncle Abe showed me was to electrify a silk tassel—he did this by stroking it with a piece of rubber—so that its threads, now charged with electricity, repelled one another and flew apart. But as soon as he brought the radium near, the threads collapsed, their electricity discharged. This was because radioactivity made the air conducting, he said, so the tassel could not hold its charge anymore. An extremely refined form of this was the gold-leaf electroscope in his lab—a sturdy jar with a metal rod through its stopper to conduct a charge and two tiny gold foil leaves suspended from this. When the electroscope was charged, the gold leaves would fly apart just like the threads of the tassel. But if one brought a radioactive substance near the jar, it would immediately discharge, and the leaves would drop. The sensitivity of the electroscope to radium was amazing—it could detect a thousand-millionth of a grain, millions of times less than the amount one could detect chemically, and it was thousands of times more sensitive even than a spectroscope.

I liked to watch Uncle Abe's radium clock, which was basically a gold-leaf electroscope with a little radium inside, in a separate, thin-walled glass vessel. The radium, emitting negative particles, would gradually get positively charged, and the gold leaves would start to diverge—until they hit the side of the vessel and got discharged; then the cycle would start all over again. This "clock" had been opening and closing its gold leaves, every three minutes, for more than thirty years, and it would go on doing so for a thousand years or more—it was the closest thing, Uncle Abe said, to a perpetual motion machine.

What had been a mild puzzle with uranium had become a much more acute one with the isolation of radium, a million times more radioactive. While uranium could darken a photographic plate (though this took several days) or discharge an ultrasensitive gold-leaf electroscope, radium did this in a fraction of a second; it glowed spontaneously with the fury of its own activity; and, as became increasingly evident in the new century, it could penetrate opaque materials, ozonize the air, tint glass, induce fluorescence, and burn and destroy the living tissues of the body, in a way that could be either therapeutic or destructive.

With radiation of every other sort, going all the way from X-rays to radio waves, energy had to be provided by an external source; but radioactive elements, apparently, had their own power and could emit energy without decrement for months or years, and neither heat nor cold nor pressure nor magnetic fields nor irradiation nor chemical reagents made the least difference to this.

Where did this immense amount of energy come from? The firmest principles in the physical sciences were the principles of conservation—that matter and energy could neither be created nor destroyed. There had never been any serious suggestion that these principles could ever be violated, and yet radium at first appeared to do exactly that—to be a *perpetuum mobile,* a free lunch, a steady and inexhaustible source of energy.

One escape from this quandary was to suppose that the energy of radioactive substances had an exterior source; this indeed was what Becquerel first suggested, on the analogy of phosphorescence—that radioactive substances absorbed energy from something, from somewhere, and then reemitted it, slowly, in their own way. (He coined the term *hyperphosphorescence* for this.)

Notions of an outside source—perhaps an X-ray-like radiation bathing the earth—had been entertained briefly by the Curies, and they had sent a sample of a radium concentrate to Hans Geitel and Julius Elster in Germany. Elster and Geitel were close friends (they were known as "the Castor and Pollux of physics"), and they were brilliant investigators, who had already shown radioactivity to be unaffected by vacua, cathode rays, or sunlight. When they took the sample down a thousand-foot mine in the Harz Mountains—a place where no X-rays could reach—they found its radioactivity undiminished.

Could radium's energy be coming from the Ether, that mysterious, immaterial medium that was supposed to fill every nook and cranny of the universe and allow for the propagation of light and gravity and all other forms of cosmic energy? This was Mendeleev's opinion when he visited the Curies, though given a special chemical twist by him, for he conceived that the Ether was composed of a very light "ether element," an inert gas able to penetrate all matter without chemical reaction, and with an atomic weight about half that of hydrogen. (This new element, he thought, had already been observed in the solar corona, and named coronium.) Beyond this, Mendeleev conceived of an ultralight etheric element, with an atomic weight less than a billionth that of hydrogen, that permeated the cosmos. Atoms of these etheric elements, he felt, attracted to the heavy atoms of uranium and thorium, and absorbed by them somehow, endowed them with their own etheric energy.[3]

(I was puzzled when I first came across reference to the Ether—often spelled *Aether,* and capitalized—confusing this with the inflammable, mobile, sharp-smelling liquid my mother kept in her anesthetic bag. A "luminiferous" Ether had been postulated by Newton as the medium in which light waves were propagated, but, as Uncle Abe told me, even in his youth people had already become suspicious of its existence. Maxwell was able to bypass it in his equations, and a famous experiment in the early 1890s had failed to show any "Ether drift," any effect of the earth's motion on the velocity of light, such as one might expect if an Ether existed. But clearly the idea of the Ether was still very strong in the minds of many scientists at the time when radioactivity was discovered, and it was natural that they should turn to it first for an explanation of its mysterious energies.[4]

---

[3] Retaining his flexibility of mind to the last, Mendeleev renounced his Etheric hypothesis the year before he died, and acknowledged his acceptance of the "unthinkable"—transmutation—as the source of radioactive energy.

[4] The Ether was pressed into many other uses, too. For Oliver Lodge, writing in 1924, it was still the needed medium for electromagnetic waves and gravitation, even though the theory of relativity, by this time, was widely known. It was also, for Lodge, the medium that provided a continuum, a matrix in which discrete particles, atoms and electrons, could be embedded. Finally, for him (as for J. J. Thomson and many others), the Ether took on a religious or metaphysical role, too—it became the medium, the realm, where spirits and Mind-at-large dwelled, where the life force of the dead maintained a sort of quasi-existence (and could perhaps be summoned forth by the efforts

But if it was imaginable—just—that a slow dribble of energy such as uranium emitted might come from an outside source, such a notion became harder to believe when faced with radium, which (as Pierre Curie and Albert Laborde would show, in 1903) was capable of raising its own weight of water from freezing to boiling in an hour.[5] It was harder still when faced with even more intensely radioactive substances, such as pure polonium (a small piece of which would spontaneously become red-hot) or radon, which was 200,000 times more radioactive than radium itself—so radioactive that a pint of it would instantly vaporize any vessel in which it was contained. Such a power to heat was unintelligible with any etheric or cosmic hypothesis.

With no plausible external source of energy, the Curies were forced to return to their original thought that the energy of radium had to have an *internal* origin, to be an "atomic property"—although a basis for this was hardly imaginable. As early as 1898, Marie Curie added a bolder, even outrageous thought, that radioactivity might come from the disintegration of atoms, that it could be "an emission of matter accompanied by a loss of weight of the radioactive substances"—a hypothesis even more bizarre, it might have seemed, than its alternatives, for it had been axiomatic in science, a fundamental assumption, that atoms were indestructible, immutable, unsplittable—the whole of

---

of mediums). Thomson and many other physicists of his generation became active members, founders, of the British Psychical Society, a reaction, perhaps, against the materialism of the time and the perceived or imagined death of God.

[5] After reading about this, I wondered whether any radioactive substances actually felt warm to the touch. I had small bars of uranium and thorium, but they felt as cool as any other metal bars. I once held Uncle Abe's little tube, with its ten milligrams of radium bromide, in my hand, but the radium was no bigger than a grain of salt, and I felt no warmth through the glass.

I was fascinated to learn from Jeremy Bernstein that he once held a sphere of plutonium in his hands—the core of an atomic bomb, no less—and found it uncannily warm to the touch.

chemistry and classical physics was built on this faith. In Maxwell's words:

> Though in the course of ages catastrophes have occurred and may yet occur in the heavens, though ancient systems may be dissolved and new systems evolved out of their ruins, the [atoms] out of which these systems are built—the foundation stones of the material universe—remain unbroken and unworn. They continue to this day as they were created—perfect in number and measure and weight.

All scientific tradition, from Democritus to Dalton, from Lucretius to Maxwell, insisted upon this principle, and one can readily understand how, after her first bold thoughts about atomic disintegration, Marie Curie withdrew from the idea, and (using unusually poetic language) ended her thesis on radium by saying, "the cause of this spontaneous radiation remains a mystery . . . a profound and wonderful enigma."

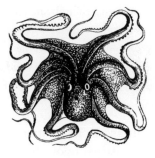

22

CANNERY ROW

The summer after the war, we went to Switzerland, because this was the only country on the Continent that had not been ravaged by war, and we longed for normality, after six years of bombing and rationing and austerity and constriction. The transformation was evident as soon as we crossed the border—the uniforms of the Swiss customs officers were new and shining, unlike the shabby uniforms on the French side. The train itself seemed to become cleaner and brighter, to move with a new efficiency and speed. Arriving in Lucerne, we were met by an electric brougham. Tall, upright, with huge plate-glass windows, a vehicle such as my parents had seen, but never ridden, in their own childhood, the ancient brougham conveyed us noiselessly to the Schweizerhof Hotel, a hotel vaster, more splendid, than anything I had ever imagined. My parents would generally choose relatively modest lodgings, but this time their instincts led them in the opposite direction, to the most sumptuous, most luxurious, most opulent hotel in Lucerne—an extravagance permitted, they felt, after six years of war.

The Schweizerhof stays in my mind for another reason, because it was here that I gave the first (and last) concert of my life. It had been a little over a year since Mrs. Silver, my piano teacher, had died, a year in which I had not touched a piano, but now something sunny, something liberating, brought me out, made me want to play again, all of a sudden, and for other people. Though I had been brought up on Bach and Scarlatti, I had grown (under Mrs. Silver's influence) to love the Romantics—especially Schumann and the propulsive, exuberant Chopin mazurkas. Many of these were technically beyond me, but I knew them, nonetheless, all fifty-odd of them, by heart, and could at least (I flattered myself) give a sense of their feel and vitality. They were miniatures, but each seemed to contain an entire world.

Somehow my parents persuaded the hotel to arrange a concert in its salon, to let me use the grand piano (it was bigger than any I had ever seen, a Bösendorfer with some extra keys our Bechstein did not have), and to announce that, on the coming Thursday night, there would be a recital by "the young English pianist Oliver Sacks." This terrified me, and I grew more and more nervous as the day approached. But when the evening came, I donned my best suit (it had been made for my bar mitzvah the month before), entered the salon, bowed, arranged my features into a smile, and (almost incontinent with terror) sat down at the piano. After the opening bars of the first mazurka, I got swept away by it, and carried it to a flamboyant conclusion. There was clapping, there were smiles, there was forgiveness of my blunders, so I charged on to the next, and the next, finishing up finally with a posthumous opus (which I vaguely imagined had somehow been completed after Chopin's death).

There was a special, rare pleasure about this performance. My chemistry and mineralogy and science were all private, shared with my uncles but with nobody else. The recital, in contrast, was open and public, with appreciation, exchange, giving and receiving. It was the opening of something new, the start of an intercourse.

We gloried shamelessly in the luxury of the Schweizerhof, lying for hours, it seemed, in the enormous marble baths, eating ourselves sick in the opulent restaurant. But eventually we grew tired of overindulgence and started to wander through the old city with its crooked streets and its sudden views of mountain and lake. We took the funicular train up its cogwheel track to the summit of Mount Rigi—my first time on a funicular, or a mountain. And then we moved to the alpine village of Arosa, where the air was cool and dry, and I saw edelweiss and gentian for the first time, and tiny churches of painted wood, and heard the alpenhorn resound from valley to valley. It was in Arosa, I think, even more than Lucerne, that a sudden sense of joyousness finally overcame me, a feeling of liberation and release, a sense of the sweetness of life, of a future, of promise. I was thirteen—thirteen!—did not life stand before me?

On the return journey, we stopped in Zurich (the town, Uncle Abe once told me, where Euler the mathematician had been born). And this stay, while otherwise unremarkable, remains in my memory for a very special reason. My father, who always sought out a swimming pool wherever he stayed, located a large municipal pool in the city. He immediately started lapping the pool, with the powerful overarm of which he was a master, but I, in a lazier mood, found a corkboard, hoisted myself upon it, and decided, for once, to let it buoy me, and just float. I lost all sense of time as I floated, lying still on the board, or paddling very gently. A strange ease, a sort of rapture came upon me—a feeling I had sometimes known in dreams. I had floated on corkboards, or rubber rings, or waterwings before, but this time something magical was happening, a slowly swelling, enormous wave of joy that lifted me higher and higher, seemed to go on and on, forever, and then finally subsided in a languorous bliss. It was the most beautiful, peaceful feeling I had ever had.

It was only when I came to take off my swimming trunks that I realized I must have had an orgasm. It did not occur to me to connect this with "sex," or other people; I did not feel anxious or guilty—but I kept it to myself, feeling it as magic, private, a benison or grace that had come upon me spontaneously, unsought. I felt as if I had discovered a great secret.

In January 1946 I moved from my prep school in Hampstead, The Hall, to a much larger school, St. Paul's, in Hammersmith. It was here, in the Walker Library, that I met Jonathan Miller for the first time: I was hidden in a corner, reading a nineteenth-century book on electrostatics—reading, for some reason, about "electric eggs"—when a shadow fell across the page. I looked up and saw an astonishingly tall, gangling boy with a very mobile face, brilliant, impish eyes, and an exuberant mop of reddish hair. We got talking together, and have been close friends ever since.

Prior to this time, I had had only one real friend, Eric Korn, whom I had known almost from birth. Eric followed me from The Hall to St. Paul's a year later, and now he and Jonathan and I formed an inseparable trio, bound not only by personal but by family bonds too (our fathers, thirty years earlier, had all been medical students together, and our families had remained close). Jonathan and Eric did not really share my love of chemistry—although they joined in the sodium-throwing experiment and one or two others—but they were intensely interested in biology, and it was inevitable, when the time came, that we would find ourselves together in the same biology class, and that all of us would fall in love with our biology teacher, Sid Pask.

Sid was a splendid teacher. He was also narrow-minded, bigoted, cursed with a hideous stutter (which we would imitate endlessly), and by no means exceptionally intelligent. By dissuasion, irony, ridicule, or force, Mr. Pask would turn us away from all other activities—from sport and sex, from religion and families, and from all our other subjects at school. He demanded that we be as single-minded as himself.

The majority of his pupils found him an impossibly demanding and exacting taskmaster. They would do all they could to escape from this pedant's petty tyranny, as they regarded it. The struggle would go on for a while, and then suddenly there was no longer any resistance—they were free. Pask no longer carped at them, no longer made ridiculous demands upon their time and energy.

Yet some of us, each year, responded to Pask's challenge. In return he gave us all of himself—all his time, all his dedication, for biology. We would stay late in the evening with him in the Natural History Museum (I once hid myself in a gallery and managed to spend the night there). We would sacrifice every weekend to plant-collecting expeditions. We would get up at dawn on freezing winter days to go on his January freshwater course. And once a year—there is still an almost intolerable sweetness about the memory—we would go with him to Millport for three weeks of marine biology.

Millport, off the western coast of Scotland, had a beautifully equipped marine biology station, where we were always given a friendly welcome and inducted into whatever experiments were going on. (Fundamental observations were being made on the development of sea urchins at this time, and Lord Rothschild was endlessly patient with the enthusiastic schoolboys who crowded around and peered into his petri dishes with the transparent pluteus larvae.) Jonathan, Eric, and I made several transects on the rocky shore together, counting all the animals and seaweeds we could on successive square-foot portions from the lichen-covered summit of the rock (*Xanthoria parietina* was the euphonious name of this lichen) to the shoreline and tidal pools below. Eric was particularly and wittily ingenious, and once, when we needed a plumb line to give us a true vertical, but did not know how to suspend it, he pried a limpet from the base of a rock, placed the tip of the plumb line beneath it, and firmly reattached it at the top as a natural drawing pin.

We all adopted particular zoological groups: Eric became enamored of sea cucumbers, holothurians; Jonathan of iridescent bristled worms, polychaetes; and I of squids and cuttlefish, octopuses, cephalopods—the most intelligent and, to my eyes, the most beautiful of invertebrates. One day we all went down to the seashore, to Hythe in Kent, where Jonathan's parents had taken a house for the summer, and went out for a day's fishing on a commercial trawler. The fishermen would usually throw back the cuttlefish that ended up in their nets (they were not popular eating in England). But I, fanatically, insisted that they keep them for me, and there must have been dozens of them on the deck by the time we came in. We took all the cuttlefish back to the house in pails and tubs, put them in large jars in the basement, and added a little alcohol to preserve them. Jonathan's parents were away, so we did not hesitate. We would be able to take all the cuttlefish back to school, to Sid—we imagined his astonished smile as we brought them in—and there would be a cuttlefish apiece for everyone in the class to dissect, two or three apiece for the cephalopod enthusiasts. I myself would give a little talk about them at the Field Club, dilating on their intelligence, their large brains, their eyes with erect retinas, their rapidly changing colors.

A few days later, the day Jonathan's parents were due to return, we heard dull thuds emanating from the basement, and going down to investigate, we encountered a grotesque scene: the cuttlefish, insufficiently preserved, had putrefied and fermented, and the gases produced had exploded the jars and blown great lumps of cuttlefish all over the walls and floor; there were even shreds of cuttlefish stuck to the ceiling. The intense smell of putrefaction was awful beyond imagination. We did our best to scrape off the walls and remove the exploded, impacted lumps of cuttlefish. We hosed down the basement, gagging, but the stench was not to be removed, and when we opened windows and doors to air out the basement, it extended outside the house as a sort of miasma for fifty yards in every direction.

Eric, always ingenious, suggested we mask the smell, or replace it, by an even stronger, but pleasant smell—a coconut essence, we decided, would fill the bill. We pooled our resources and bought a large bottle of this, which we used to douche the basement, and then distributed liberally through the rest of the house and its grounds.

Jonathan's parents arrived an hour later and, advancing toward the house, hit an overwhelming scent of coconut. But as they drew nearer they hit a zone dominated by the stench of putrefied cuttlefish—the two smells, the two vapors, for some curious reason, had organized themselves in alternating zones about five or six feet wide. By the time they reached the scene of our accident, our crime, the basement, the smell was insupportable for more than a few seconds. The three of us were all in deep disgrace over the incident, I especially, since it had arisen from my greed in the first place (would not a single cuttlefish have done?) and my folly in not realizing how much alcohol they would need. Jonathan's parents had to cut short their holiday and leave the house (the house itself, we heard, remained uninhabitable for months). But my love of cuttlefish remained unimpaired.

Perhaps there was a chemical reason for this, as well as a biological one, for cuttlefish (like many other molluscs and crustaceans) had blue blood, not red, because they had evolved a completely different system for transporting oxygen from the one we vertebrates had. Whereas our red respiratory pigment, hemoglobin, contained iron, their bluish green pigment, hemocyanin, contained copper. Both iron and copper had excellent reduction potential: they could easily take up oxygen, moving to a higher oxidation state, and then relinquish it, get reduced, as needed. I wondered if their neighbors in the periodic table (some with even greater redox potential) were ever exploited as respiratory pigments, and got most excited when I heard that some sea squirts, tunicates, were extremely rich in the element vanadium, and had special cells, vanadocytes, devoted to storing it. Why they contained these was a mystery; they did not seem to be part of an oxygen-transport system. Absurdly, impudently, I thought I might solve this mystery during one of our annual excursions to Millport. But I got no further than collecting a bushel of sea squirts (with the same greed, the same inordinacy, that had caused me to collect too many cuttlefish). I could incinerate these, I thought, and measure the vanadium content of their ash (I had read that this could exceed 40 percent in some species). And this gave me the only commercial idea I have ever had: to open a vanadium farm—acres of sea meadows, seeded with sea squirts. I would get them to extract the precious vanadium from seawater, as they had been doing very efficiently for the last 300 million years, and then sell it for £500 a ton. The only problem, I realized, aghast at my own genocidal thoughts, would be the veritable holocaust of sea squirts required.

The organic, with all its complexities, was entering my own life, transforming me, in the strongholds of my own body. Suddenly I started to grow very fast; hair sprouted on my face, in my armpits, around my genitals; and my voice—still a clear treble when I chanted my haftorah—now started to break, to change pitch erratically. In biology class at school, I developed a sudden, intense interest in the reproductive systems of animals and plants, "lower" ones particularly, invertebrates and gymnosperms. The sexuality of cycads, of ginkgos, intrigued me, the way they preserved still-motile spermatozoa, like ferns, but had such large and well-protected seeds. And cephalopods, squid, were even more interesting, for the males actually thrust a modified arm bearing spermatophores into the mantle cavity of the female. I was still at a great distance from human sexuality, my own sexuality, but I started to find sexuality as a subject extremely intriguing, almost as interesting, in its way, as valency or periodicity.

But enamored though we were with biology, none of us could be as monomaniacal as Mr. Pask. There were all the pulls of youth, of adolescence, and all the energy of minds that wanted to explore in all directions, not yet ready to be committed.

My own mood had been predominantly scientific for four years; a passion for order, for formal beauty, had drawn me on— the beauty of the periodic table, the beauty of Dalton's atoms. Bohr's quantal atom seemed to me a heavenly thing, groomed, as it were, to last for an eternity. At times I felt a sort of ecstasy at the formal intellectual beauty of the universe. But now, with the onset of other interests, I sometimes felt the opposite of this, a sort of emptiness or aridity inside, for the beauty, the love of science, no longer entirely satisfied me, and I hungered now for the human, the personal.

It was music especially which brought this hunger out, and assuaged it; music which made me tremble, or want to weep, or howl; music which seemed to penetrate me to the core, to call to my condition—even though I could not say what it was "about," why it affected me the way it did. Mozart, above all, raised feelings of an almost unbearable intensity, but to define these feelings was beyond me, perhaps beyond the power of language itself.

Poetry became important in a new, personal way. We had "done" Milton and Pope at school, but now I started to discover them for myself. There were lines in Pope of an overwhelming tenderness—"Die of a rose in aromatic pain"—which I would whisper to myself again and again, until they transported me to another world.

Jonathan and Eric and I had all grown up with a love of reading and literature: Jonathan's mother was a novelist and biographer, and Eric, the most precocious of us, had been reading poetry since he was eight. My own reading tended more to history and biography, and especially personal narratives and journals. (I had also started keeping a journal of my own at this point.) My own tastes being (as they saw it) somewhat restricted, Eric and Jonathan introduced me to a wider range of writing—Jonathan to Selma Lagerlöf and Proust (I had only heard of Joseph-Louis Proust, the chemist, not of Marcel), and Eric to T. S. Eliot, whose poetry, he contended, was greater than Shakespeare's. And it was Eric who took me to the Cosmo Restaurant in Finchley Road, where over lemon tea and strudel we would listen to a young medical-student poet, Dannie Abse, recite the poems he had just written.

The three of us decided, impudently, to form a Literary Society at school; one already existed, it was true—the Milton Society—but it had been moribund for many years. Jonathan was to be our secretary, Eric our treasurer, and I (though I felt I was the most ignorant of the three, as well as the shyest) its president.

We announced a first meeting to explore things, and a curious group came. There was a strong desire to invite outside speakers to address us—poets, playwrights, novelists, journalists—and it fell upon me, as president, to tempt them into coming. An astonishing number of writers did come to our meetings— drawn (I imagine) by the sheer eccentricity of the invitations, their absurd mixture of childishness and grown-upness, and the idea, perhaps, of a crowd of enthusiastic boys who had actually read some of their works, and who were agog to meet them. The biggest coup would have been Bernard Shaw—but he sent me a charming postcard, in a shaky hand, saying that though he would love to come, he was too old to travel (he was ninety-three and three-quarters, he wrote). With our invited speakers, and the vehement discussions that followed, we became very popular, and fifty or seventy boys would sometimes turn up for our weekly meetings, far more than had ever been seen at the sedate meetings of the Milton Society. In addition, we published a smudgy, purple-inked mimeographed journal, the *Prickly Pear,* which included pieces by the students and occasionally one of the masters and, very occasionally, from "real" outside writers.

But our very success, and perhaps other, never explicitly avowed thoughts—that we were mocking authority, that we had subversive intent, that we had "killed" the Milton Society (which had now, in reaction, suspended its never-frequent meetings), and that we were a lot of obnoxious, noisy, clever Jewish boys who needed to be put down—led to our demise. The High Master called me in one day, and said without ceremony, "Sacks, you're dissolved."

"What do you mean, sir?" I stammered. "You can't just 'dissolve' us."

"Sacks, I can do whatever I want. Your literary society is dissolved as of this moment."

"But why, sir?" I asked. "What are your reasons?"

"I don't have to give them to you, Sacks. I don't have to have reasons. You can go now, Sacks. You don't exist. You don't exist anymore." With this, he snapped his fingers—a gesture of dismissal, of annihilation—and went back to his work.

I took the news to Eric and Jonathan, and to others who had been members of our society. We were outraged, and puzzled, but we felt completely helpless. The High Master had authority, absolute power, and there was nothing we could do to resist it or oppose him.

Cannery Row was published in 1945 or 1946, and I must have read it fairly soon after—perhaps in 1948, when I was doing biology in school, and marine biology had been added to my list of interests. I loved the figure of Doc, his searching for baby octopus in the tidal pools near Monterey, his drinking beer milkshakes with the boys, the idyllic ease and sweetness of his life. I thought that I, too, would like to have a life like him, to live in magical, mythical California (already, with cowboy films, a land of fantasy for me). America was increasingly in my thoughts as I entered my teens—it had been our great ally in the war; its power, its resources, were almost unlimited. Had it not made the world's first atomic bomb? American soldiers on leave walked the streets of London—their gestures, their speech, seeming to emit a self-confidence, a nonchalance, an ease almost unimaginable to us after six years of war. *Life* magazine, in its large spreads, pictured mountains, canyons, deserts, landscapes of a spaciousness and magnificence beyond anything in Europe, along with American towns full of smiling, eager, well-nourished people, their houses gleaming, their shops full, enjoying a life of plenty and gaiety unimaginable to us, with the tight rationing, the pinched consciousness of the war years still upon us. To this glamorous picture of transatlantic ease, and bigger-than-life spontaneity and splendor, musicals like *Annie Get Your Gun* and *Oklahoma!* added a further mythopoeic force. It was in this atmosphere of romantic enlargement that *Cannery Row* and (despite its sickliness) its sequel, *Sweet Thursday,* had such an impact on me.

If I had (in my St. Lawrence days) sometimes imagined a mythical past, I now started to have fantasies of the future, to imagine myself as a scientist or naturalist on the coasts or in the great outback of America. I read accounts of Lewis and Clark's journey, I read Emerson and Thoreau, and above all, I read John Muir. I fell in love with the sublime and romantic landscapes of Albert Bierstadt and the beautiful, sensuous photographs of Ansel Adams (I had fantasies, on occasion, of becoming a landscape photographer myself).

When I was sixteen or seventeen, deeply in love now with marine biology, I wrote to marine biology laboratories all over the States—to Woods Hole in Massachusetts, to the Scripps Institution in La Jolla, to the Golden Gate Aquarium in San Francisco, and of course to Cannery Row in Monterey (by this time I knew that "Doc" was a real person, Ed Ricketts). I got affable replies, I think, from them all, welcoming my interest and enthusiasm, but also indicating very clearly that I needed some real qualifications, too, and that I should think about recontacting them when I had a degree in biology (when I eventually made it to California, ten years later, it was not as a marine biologist, but as a neurologist).

23

$$\left(\text{226}\right) = \left(\text{222}\right) + \left(\text{4}\right)$$

Radium. Emanation. Helium

THE WORLD SET FREE

The Curies had noticed from the start that their radioactive substances showed a strange power to "induce" radioactivity all around them. They found this both intriguing and irritating, for the contamination of their equipment made it nearly impossible to measure the radioactivity of the samples themselves:

> The different objects used in the chemical laboratory [Marie wrote in her thesis] . . . soon acquire radioactivity. Dust particles, the air of the room, clothing, all become radioactive. The air of the room becomes a conductor. In our laboratory the evil has become acute, and we no longer have any apparatus properly insulated.[1]

I thought of our own house and of Uncle Abe's house as I read this passage, wondering whether they, too, in their mild way, had become radioactive—whether the radium-painted dials of Uncle

---

[1] Marie Curie's own laboratory notebooks, a century later, are still considered too dangerous to handle and are kept in lead-lined boxes.

Abe's clocks were inducing radioactivity in everything around them and filling the air, silently, with penetrating rays.

The Curies (like Becquerel) were at first inclined to attribute this "induced radioactivity" to something immaterial, or to see it as a "resonance," perhaps analogous to phosphorescence or fluorescence. But there were also indications of a material emission. They had found, as early as 1897, that if thorium was kept in a tightly shut bottle its radioactivity increased, returning to its previous level as soon as the bottle was opened. But they did not follow up on this observation, and it was Ernest Rutherford who first realized the extraordinary implication of this: that a new substance was coming into being, being generated by the thorium; a far more radioactive substance than its parent.

Rutherford enlisted the help of the young chemist Frederick Soddy, and they were able to show that the "emanation" of thorium was in fact a material substance, a gas, which could be isolated. It could be liquefied, almost as easily as chlorine, but it did not react with any chemical reagent; it was in fact just as inert as argon and the other newly discovered inert gases. At this point Soddy thought that the "emanation" of thorium might *be* argon, and he was (as he wrote later)

> overwhelmed with something greater than joy — I cannot very well express it — a kind of exaltation. . . . I remember quite well standing there transfixed as though stunned by the colossal impact of the thing and blurting out — or so it seemed at the time, "Rutherford, this is transmutation: the thorium is disintegrating and transmuting itself into argon gas."
>
> Rutherford's reply was typically aware of more practical implications: "For Mike's sake, Soddy, don't call it *transmutation*. They'll have our heads off as alchemists."

But the new gas was not argon; it was a brand-new element with its own unique bright-line spectrum. It diffused very slowly and was exceedingly dense — 111 times as dense as hydrogen, whereas argon was only 20 times as dense. Assuming a molecule of the new gas was monatomic, containing only one atom like the other inert gases, its atomic weight would be 222. Thus it was the heaviest and last in the inert gas series, and as such could take its place in the periodic table, as the final member of Mendeleev's Group 0. Rutherford and Soddy provisionally named it thoron or Emanation.

Thoron disappeared with great speed—half of it was gone in a minute, three-quarters in two minutes, and in ten minutes it was no longer detectable. It was the rapidity of this breakdown (and the appearance of a radioactive deposit in its place) which allowed Rutherford and Soddy to perceive what had not been clear with uranium or radium—that there was indeed a continuous disintegration of the atoms of radioactive elements, and with this their transformation to other atoms.

Each radioactive element, they found, had its own characteristic rate of breakdown, its own "half-life." The half-life of an element could be given with extraordinary precision, so that the half-life of one radon isotope, for instance, could be calculated as 3.8235 days. But the life of an individual atom could not be predicted in the least. I became more and more bewildered by this thought, and kept rereading Soddy's account:

> The chance at any instant whether an atom disintegrates or not in any particular second is *fixed*. It has nothing to do with any external or internal consideration we know of, and in particular is *not* increased by the fact that the atom has already survived any period of past time. . . . All that can be said is that the immediate cause of atomic disintegration appears to be due to chance.

The life span of an individual atom, apparently, might vary from zero to infinity, and there was nothing to distinguish an atom "ready" to disintegrate from one that still had a billion years before it.

I found this profoundly mystifying and disconcerting, that an atom might disintegrate at any time, without any "reason" to do so. It seemed to remove radioactivity from the realm of continuity or process, from the intelligible, causal universe—and to hint at a realm where laws of the classical sort meant nothing whatsoever.

The half-life of radium was much longer than that of its emanation, radon—about 1,600 years. But this was still very small compared to the age of the earth—why, then, if it steadily decayed, had all the earth's radium not disappeared long ago? The answer, Rutherford inferred, and was soon able to demonstrate, was that radium itself was produced by elements with a much longer half-life, a whole train of substances that he could trace back to the parent element, uranium. Uranium in turn had a half-life of four and a half billion years, roughly the age of the earth itself. Other cascades of radioactive elements were derived from thorium, which had an even longer half-life than uranium. Thus the earth was still living, in terms of atomic energy, on the uranium and thorium that had been present when the earth formed.

These discoveries had a crucial impact on a long-standing debate about the age of the earth. The great physicist Kelvin, writing in the early 1860s, soon after the publication of *The Origin of Species,* had argued that, based on its rate of cooling, and assuming no source of heat other than the sun, the earth could be no more than twenty million years old, and that in another five million years it would become too cold to support life. This calculation was not only dismaying in itself, but was impossible to reconcile with the fossil record, which indicated that life had been present for hundreds of millions of years—and yet there seemed no way of rebutting it. Darwin was greatly disturbed by this.

It was only with the discovery of radioactivity that the conundrum was solved. The young Rutherford, it was said, nervously facing the famous Lord Kelvin, now eighty years old, suggested that Kelvin's calculation had been based on a false assumption. There *was* another source of warmth besides the sun, Rutherford said, and a very important one for the earth. Radioactive elements (chiefly uranium and thorium, and their breakdown products, but also a radioactive isotope of potassium) had served to keep the earth warm for billions of years and to protect it from the premature heat-death that Kelvin had predicted. Rutherford held up a piece of pitchblende, the age of which he had estimated from the amount of helium it contained. *This* piece of the earth, he said, was at least 500 million years old.

Rutherford and Soddy were ultimately able to delineate three separate radioactive cascades, each containing a dozen or so breakdown products emanating from the disintegration of the original parent elements. Could all of these breakdown products be different elements? There was no room in the periodic table for three dozen elements between bismuth and thorium—room for half a dozen, perhaps, but not much more. Only gradually did it become clear that many of the elements were just versions of one another; the emanations of radium and thorium and actinium, for example, though they had widely differing half-lives, were chemically identical, all the same element, though with slightly different atomic weights. (Soddy later named these isotopes.) And the end points of each series were similar—radium G, actinium E, and thorium E, so-called, were all isotopes of lead.

Every substance in these cascades of radioactivity had its own unique radio signature, a half-life of fixed and invariable duration, as well as a characteristic radiation emission, and it was this which allowed Rutherford and Soddy to sort them all out, and in so doing to found the new science of radiochemistry.

The idea of atomic disintegration, first raised and then retreated from by Marie Curie, could no longer be denied. It was evident that every radioactive substance disintegrated in the act of giving off energy and turned into another element, that transmutation lay at the heart of radioactivity.

I loved chemistry in part because it was a science of transformations, of innumerable compounds based on a few dozen elements, themselves fixed and invariant and eternal. The feeling of the elements' stability and invariance was crucial to me psychologically, for I felt them as fixed points, as anchors, in an unstable world. But now, with radioactivity, came transformations of the most incredible sort. What chemist would have conceived that out of uranium, a hard, tungsteny metal, there could come an alkaline earth metal like radium; an inert gas like radon; a tellurium-like element, polonium; radioactive forms of bismuth and thallium; and, finally, lead—exemplars of almost every group in the periodic table?

No chemist would have conceived this (though an alchemist might), because the transformations lay beyond the sphere of chemistry. No chemical process, no chemical attack, could ever alter the identity of an element, and this applied to the radioactive elements too. Radium, chemically, behaved similarly to barium; its radioactivity was a different property altogether, wholly unrelated to its chemical or physical properties. Radioactivity was a marvelous (or terrible) addition to these, a wholly other property (and one that annoyed me at times, for I loved the tungstenlike density of metallic uranium, and the fluorescence and beauty of its minerals and salts, but I felt I could not handle them safely for long; similarly I was infuriated by the intense radioactivity of radon, which otherwise would have made an ideal heavy gas).

Radioactivity did not alter the realities of chemistry, or the notion of elements; it did not shake the idea of their stability and identity. What it did do was to hint at two realms in the atom— a relatively superficial and accessible realm governing chemical reactivity and combination, and a deeper realm, inaccessible to all the usual chemical and physical agents and their relatively small energies, where any change produced a fundamental alteration of the element's identity.

Uncle Abe had in his house a "spinthariscope," just like the ones advertised on the cover of Marie Curie's thesis. It was a beautifully simple instrument, consisting of a fluorescent screen and a magnifying eyepiece, and inside, an infinitesimal speck of radium. Looking through the eyepiece, one could see dozens of scintillations a second—when Uncle Abe handed me this, and I held it up to my eye, I found the spectacle enchanting, magical, like looking at an endless display of meteors or shooting stars.

Spinthariscopes, at a few shillings each, were fashionable scientific toys in Edwardian drawing rooms—a new and uniquely twentieth-century accession, next to the stereoscopes and Geissler tubes inherited from Victorian times. But if they made their appearance as a sort of toy, it was rapidly appreciated that they also showed one something fundamentally important, for the tiny sparks or scintillations one saw came from the disintegration of individual atoms of radium, from the individual alpha particles each shot off as it exploded. No one would have imagined, Uncle Abe said, that we would ever be able to see the effects of individual atoms, much less to count them individually.

"Here there is less than a millionth of a milligram of radium, and yet, on the small area of the screen, there are dozens of scintillations a second. Imagine how many there would be if we had a gram of radium—a thousand million times this amount."

"A hundred thousand million," I calculated.

"Close," Uncle said. "One hundred and thirty-six thousand million, to be exact—the number never varies. Every second, one hundred and thirty-six thousand million atoms in a gram of radium disintegrate, shoot off their alpha particles—and if you think of this going on for thousands of years, you'll get some idea of how many atoms there are in a single gram of radium."

Experiments around the turn of the century had shown that

not only alpha rays but several other sorts of ray were being emitted by radium. Most of the phenomena of radioactivity could be attributed to these different sorts of rays: the ability to ionize air was especially the prerogative of the alpha rays, while the ability to elicit fluorescence or affect photographic plates was more marked with the beta rays. Every radioactive element had its own characteristic emissions: thus radium preparations emitted both alpha and beta rays, where polonium preparations emitted only alpha rays. Uranium affected a photographic plate more quickly than thorium, but thorium was more potent in discharging an electroscope.

The alpha particles emitted by radioactive decay (they were later shown to be helium nuclei) were positively charged and relatively massive—thousands of times more massive than beta particles or electrons—and they traveled in undeviating straight lines, passing straight through matter, ignoring it, without any scattering or deflection (although they might lose some of their velocity in so doing). This, at least, appeared to be the case, though in 1906 Rutherford observed that there might be, very occasionally, small deflections. Others ignored this, but to Rutherford these observations were fraught with possible significance. Would not alpha particles be ideal projectiles, projectiles of atomic proportions, with which to bombard other atoms and sound out their structure? He asked his young assistant Hans Geiger and a student, Ernest Marsden, to set up a scintillation experiment using screens of thin metal foils, so that one could keep count of every alpha particle that bombarded these. Firing alpha particles at a piece of gold foil, they found that roughly one in eight thousand particles showed a massive deflection—of more than 90 degrees, and sometimes even 180 degrees. Rutherford was later to say, "It was quite the most incredible event that ever happened to me in my life. It was almost as incredible as if you fired a fifteen-inch shell at a piece of tissue paper and it came back and hit you."

Rutherford pondered these curious results for almost a year, and then, one day, as Geiger recorded, he "came into my room, obviously in the best of moods, and told me that now he knew what the atom looked like and what the strange scatterings signified."

Atoms, Rutherford had realized, could not be a homogenous jelly of positivity stuck with electrons like raisins (as J. J. Thomson had suggested, in his "plum pudding" model of the atom), for then the alpha particles would always go through them. Given the great energy and charge of these alpha particles, one had to assume that they had been deflected, on occasion, by something even more positively charged than themselves. Yet this happened only once in eight thousand times. The other 7,999 particles might whiz through, undeflected, as if most of the gold atoms consisted of empty space; but the eight-thousandth was stopped, flung back in its tracks, like a tennis ball hitting a globe of solid tungsten. The mass of the gold atom, Rutherford inferred, had to be concentrated at the center, in a minute space, not easy to hit—as a nucleus of almost inconceivable density. The atom, he proposed, must consist overwhelmingly of empty space, with a dense, positively charged nucleus only a hundred-thousandth its diameter, and a relatively few, negatively charged electrons in orbit about this nucleus—a miniature solar system, in effect.

Rutherford's experiments, his nuclear model of the atom, provided a structural basis for the enormous differences between radioactive and chemical processes, the millionfold differences of energy involved (Soddy would dramatize this, in his popular lectures, by holding a one-pound jar of uranium oxide aloft in one hand—this, he would say, had the energy of a hundred and sixty tons of coal).

Chemical change or ionization involved the addition or removal of an electron or two, and this required only a modest energy of two or three electron-volts, such as could be produced easily—by a chemical reaction, by heat, by light, or by a simple 3-volt battery. But radioactive processes involved the nuclei of atoms, and since these were held together by far greater forces, their disintegration could release energies of far greater magnitude—some millions of electron-volts.

Soddy coined the term *atomic energy* soon after the turn of the nineteenth century, ten years or more before the nucleus was discovered. No one had known, or been able to make a remotely plausible guess, as to how the sun and stars could radiate so much energy, and continue to do so for millions of years. Chemical energy would be ludicrously inadequate—a sun made of coal would burn itself out in ten thousand years. Could radioactivity, atomic energy, provide the answer?

> Supposing [wrote Soddy] . . . our sun . . . were made of pure radium . . . there would be no difficulty in accounting for its outpourings of energy.

Could transmutation, which occurs naturally in radioactive substances, be produced artificially, Soddy wondered.[2] He was moved by this thought to rapturous, millennial, and almost mystical heights:

> Radium has taught us that there is no limit to the amount of energy in the world. . . . A race which could transmute matter would have little need to earn its bread by the sweat of its brow. . . . Such a race could transform a desert continent, thaw the frozen poles, and make the whole world one smiling Garden of Eden. . . . An entirely new prospect has been opened up. Man's inheritance has increased, his aspirations have been uplifted, and his destiny has been ennobled to an extent beyond our present

---

[2] Soddy envisaged this artificial transmutation fifteen years before Rutherford achieved it, and imagined explosive or controlled atomic disintegrations long before fission or fusion were discovered.

power to foretell. . . . One day he will attain the power to regulate for his own purposes the primary fountains of energy which Nature now so jealously conserves for the future.

I read Soddy's book *The Interpretation of Radium* in the last year of the war, and I was enraptured by his vision of endless energy, endless light. Soddy's heady words gave me a sense of the intoxication, the sense of power and redemption, that had attended the discovery of radium and radioactivity at the start of the century.

But side by side with this, Soddy voiced the dark possibilities, too. These indeed had been in his mind almost from the start, and, as early as 1903, he had spoken of the earth as "a storehouse stuffed with explosives, inconceivably more powerful than any we know of." This note was frequently sounded in *The Interpretation of Radium,* and it was Soddy's powerful vision that inspired H. G. Wells to go back to his early science-fiction style and publish, in 1914, *The World Set Free* (Wells actually dedicated his book to *The Interpretation of Radium*). Here Wells envisaged a new radioactive element called Carolinum, whose release of energy was almost like a chain reaction:[3]

> Always before in the development of warfare the shells and rockets fired had been but momentarily explosive, they had gone off in an instant once and for all . . . but Carolinum . . . once its degenerative process had been induced, continued a furious radiation of energy and nothing could arrest it.

I thought of Soddy's prophesies, and Wells's, in August of 1945, when we heard the news of Hiroshima. My feelings about the atomic bomb were strangely mixed. Our war, after all, was over, V-E Day was past; unlike the Americans, we had not suf-

---

[3] It was reading *The World Set Free* in the 1930s that set Leo Szilard to thinking of chain reactions and getting a secret patent on these in 1936; in 1940 he persuaded Einstein to send his famous letter to Roosevelt about the possibilities of an atomic bomb.

fered Pearl Harbor, or the terrible struggles in Guam and Saipan; we had not been in direct combat with the Japanese. The atomic bombings seemed, in some ways, like a terrible postscript to the war, a hideous demonstration that perhaps did not need to be made.

And yet I also had, as many did, a sense of jubilation at the scientific achievement of splitting the atom, and I was enthralled by the Smyth Report, which came out in August of 1945 and gave a full description of the making of the bomb. The full horror of the bomb did not hit me until the following summer, when John Hersey's "Hiroshima" was published in a special one-article edition of *The New Yorker* (Einstein, it was said, bought a thousand copies of this issue) and broadcast soon after by the BBC on the Third Programme. Up to this point, chemistry and physics had been for me a source of pure delight and wonder, and I was insufficiently conscious, perhaps, of their negative powers. The atomic bombs shook me, as they did everybody. Atomic or nuclear physics, one felt, could never again move with the same innocence and lightheartedness as it had in the days of Rutherford and the Curies.

24

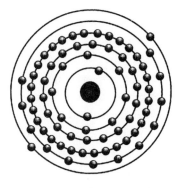

Tungsten (W⁷⁴)

BRILLIANT LIGHT

How many elements would God need to build a universe? Fifty-odd elements were known by 1815; and, if Dalton was right, this meant fifty different sorts of atom. But surely God would not need fifty different building blocks for His universe—surely He would have designed it more economically than this. William Prout, a chemically minded physician in London, observing that atomic weights were close to whole numbers and therefore multiples of the atomic weight of hydrogen, speculated that hydrogen was in fact the primordial element, and that all other elements had been built from it. Thus God needed to create only one sort of atom, and all the others, by a natural "condensation," could be generated from this.

Unfortunately, some elements turned out to have fractional

atomic weights. One could round off a weight that was slightly less or slightly more than a whole number (as Dalton did), but what could one do with chlorine, for example, with its atomic weight of 35.5? This made Prout's hypothesis difficult to maintain, and further difficulties emerged when Mendeleev made the periodic table. It was clear, for example, that tellurium came, in chemical terms, before iodine, but its atomic weight, instead of being less, was greater. These were grave difficulties, and yet throughout the nineteenth century Prout's hypothesis never really died—it was so beautiful, so simple, many chemists and physicists felt, that it must contain an essential truth.

Was there perhaps some atomic property that was more integral, more fundamental than atomic weight? This was not a question that could be addressed until one had a way of "sounding" the atom, sounding, in particular, its central portion, the nucleus. In 1913, a century after Prout, Harry Moseley, a brilliant young physicist working with Rutherford, set about exploring atoms with the just-developed technique of X-ray spectroscopy. His experimental setup was charming and boyish: using a little train, each car carrying a different element, moving inside a yard-long vacuum tube, Moseley bombarded each element with cathode rays, causing them to emit characteristic X-rays. When he came to plot the square roots of the frequencies against the atomic number of the elements, he got a straight line; and plotting it another way, he could show that the increase in frequency showed sharp, discrete steps or jumps as he passed from one element to the next. This had to reflect a fundamental atomic property, Moseley believed, and that property could only be nuclear charge.

Moseley's discovery allowed him (in Soddy's words) to "call the roll" of the elements. No gaps could be allowed in the sequence, only even, regular steps. If there was a gap, it meant that an element was missing. One now knew for certain the order of the elements, and that there were ninety-two elements and ninety-two only, from hydrogen to uranium. And it was now clear that there were seven missing elements, and seven only, still to be found. The "anomalies" that went with atomic weights were resolved: tellurium might have a slightly higher atomic weight than iodine, but it was element number 52, and iodine was 53. It was atomic number, not atomic weight, that was crucial.

The brilliance and swiftness of Moseley's work, which was all done in a few months of 1913–14, produced mixed reactions among chemists. Who was this young whippersnapper, some older chemists felt, who presumed to complete the periodic table, to foreclose the possibility of discovering any new elements other than the ones he had designated? What did he know about chemistry—or the long, arduous processes of distillation, filtration, crystallization that might be necessary to concentrate a new element or analyze a new compound? But Urbain, one of the greatest analytic chemists of all—a man who had done fifteen thousand fractional crystallizations to isolate lutecium—at once appreciated the magnitude of the achievement, and saw that far from disturbing the autonomy of chemistry, Moseley had in fact confirmed the periodic table and reestablished its centrality. "The law of Moseley . . . confirmed in a few days the conclusions of my twenty years of patient work."

Atomic numbers had been used before to denote the ordinal sequence of elements ranked by their atomic weight, but Moseley gave atomic numbers real meaning. The atomic number indicated the nuclear charge, indicated the element's identity, its chemical identity, in an absolute and certain way. There were, for example, several forms of lead—isotopes—with different atomic weights, but all of these had the same atomic number, 82. Lead was essentially, quintessentially, number 82, and it could not change its atomic number without ceasing to be lead. Tungsten was necessarily, unavoidably, element 74. But how did its 74-ness endow it with its identity?

.   .   .

Though Moseley had shown the true number and order of the elements, other fundamental questions still remained, questions that had vexed Mendeleev and the scientists of his time, questions that vexed Uncle Abe as a young man, and questions that now vexed me as the delights of chemistry and spectroscopy and playing with radioactivity gave way to a raging Why? Why? Why? Why were there elements in the first place, and why did they have the properties they did? What made the alkali metals and the halogens, in their opposite ways, so violently active? What accounted for the similarity of the rare-earth elements and the beautiful colors and magnetic properties of their salts? What generated the unique and complex spectra of the elements, and the numerical regularities which Balmer had discerned in these? What, above all, allowed the elements to be stable, to maintain themselves unchanged for billions of years, not only on the earth, but, seemingly, in the sun and stars too? These were the sorts of questions Uncle Abe had agonized about as a young man, forty years before—but in 1913, he told me, all these questions and dozens of others had, in principle, been answered and a new world of understanding had suddenly opened.

Rutherford and Moseley had chiefly been concerned with the nucleus of the atom, its mass and units of electrical charge. But it was the orbiting electrons, presumably, their organization, their bonding, that determined an element's chemical properties, and (it seemed likely) many of its physical properties, too. And here, with the electrons, Rutherford's model of the atom came to grief. According to classical, Maxwellian physics, such a solar-system atom could not work, for the electrons whirling about the nucleus more than a trillion times a second should create radiation in the form of visible light, and such an atom would emit a momentary flash of light, then collapse inward as its electrons, their energy lost, crashed into the nucleus. But the actuality (barring radioactivity) was that elements and their atoms lasted for billions of years, lasted in effect forever. How then could an atom possibly be stable, resisting what would seem to be an almost instantaneous fate?

Utterly new principles had to be invoked, or invented, to come to terms with this impossibility. Learning of this was the third ecstasy of my life, at least of my "chemical" life—the first having been learning of Dalton and atomic theory, and the second of Mendeleev and his periodic table. But the third, I think, was in some ways the most stunning of all, because it contravened (or seemed to) all the classical science I knew, and all I knew of rationality and causality.

It was Niels Bohr, also working in Rutherford's lab in 1913, who bridged the impossible, by bringing together Rutherford's atomic model with Planck's quantum theory. The notion that energy was absorbed or emitted not continuously but in discrete packets, "quanta," had lain silently, like a time bomb, since Planck had suggested it in 1900. Einstein had made use of the idea in relation to photoelectric effects, but otherwise quantum theory and its revolutionary potential had been strangely neglected, until Bohr seized on it to bypass the impossibilities of the Rutherford atom. The classical view, the solar-system model, would allow electrons an infinity of orbits, all unstable, all crashing into the nucleus. Bohr postulated, by contrast, an atom that had a limited number of discrete orbits, each with a specific energy level or quantal state. The least energetic of these, the closest to the nucleus, Bohr called the "ground state"—an electron could stay here, orbiting the nucleus, without emitting or losing any energy, forever. This was a postulate of startling, outrageous audacity, implying as it did that the classical theory of electromagnetism might be inapplicable in the minute realm of the atom.

There was, at the time, no evidence for this; it was a pure leap of inspiration, imagination—not unlike the leaps he now posited for the electrons themselves, as they jumped, without warning or intermediates, from one energy level to another. For, in addition to the electron's ground state, Bohr postulated, there were higher-energy orbits, higher-energy "stationary states," to which electrons might be briefly translocated. Thus if energy of the right frequency was absorbed by an atom, an electron could move from its ground state into a higher-energy orbit, though sooner or later it would drop back to its original ground state, emitting energy of exactly the same frequency as it had absorbed—this is what happened in fluorescence or phosphorescence, and it explained the identity of spectral emission and absorption lines, which had been a mystery for more than fifty years.

Atoms, in Bohr's vision, could not absorb or emit energy except by these quantum jumps—and the discrete lines of their spectra were simply the expression of the transitions between their stationary states. The increments between energy levels decreased with distance from the nucleus, and these intervals, Bohr calculated, corresponded exactly to the lines in the spectrum of hydrogen (and to Balmer's formula for these). This coincidence of theory and reality was Bohr's first great triumph. Einstein felt that Bohr's work was "an *enormous* achievement," and, looking back thirty-five years later, he wrote, "[it] appears to me as a miracle even today. . . . This is the highest form of musicality in the sphere of thought." The spectrum of hydrogen—spectra in general—had been as beautiful and meaningless as the markings on butterflies' wings, Bohr remarked; but now one could see that they reflected the energy states within the atom, the quantal orbits in which the electrons spun and sang. "The language of spectra," wrote the great spectroscopist Arnold Sommerfeld, "has been revealed as an atomic music of the spheres."

Could quantum theory be extended to more complex, multielectron atoms? Could it explain their chemical properties, explain the periodic table? This became Bohr's focus as scientific life resumed after the First World War.[1]

As one moved up in atomic number, as the nuclear charge or number of protons in the nucleus increased, an equal number of electrons had to be added to preserve the neutrality of the atom. But the addition of these electrons to an atom, Bohr envisaged, was hierarchical and orderly. While he had concerned himself at first with the potential orbits of hydrogen's lone electron, he now extended his notion to a hierarchy of orbits or shells for all the elements. These shells, he proposed, had definite and discrete energy levels of their own, so that if electrons were added one by one, they would first occupy the lowest-energy orbit available, and when that was full, the next-lowest orbit, then the next, and so on. Bohr's shells corresponded to Mendeleev's periods, so that the first, innermost shell, like Mendeleev's first period, accommodated two elements, and two only. Once this shell was completed, with its two electrons, a second shell began, and this, like Mendeleev's second period, could accommodate eight electrons and no more. Similarly for the third period or shell. By such a building-up, or *aufbau,* Bohr felt, all the elements could be sys-

---

[1] By 1914, the scientists of Britain and France and Germany and Austria were all caught up, in various ways, in the First World War. Pure chemistry and physics were largely suspended for the duration, and applied science, war science, took its place. Rutherford ceased his fundamental research, and his lab was reorganized for work on submarine detection. Geiger and Marsden, who had observed the alpha-particle deflections that gave rise to Rutherford's atom, found themselves at the Western Front, on different sides. Chadwick and Ellis, younger colleagues of Rutherford's, were prisoners of war in Germany. And Moseley, aged twenty-eight, was killed by a bullet in the brain, at Gallipoli. My father often used to talk of the young poets, the intellectuals, the cream of a generation wiped out tragically in the Great War. Most of the names he mentioned were unknown to me, but Moseley's was the one I knew, and the one I mourned most.

tematically constructed, and would naturally fall into their proper places in the periodic table.

Thus the position of each element in the periodic table represented the number of electrons in its atoms, and each element's reactivity and bonding could now be seen in electronic terms, in accordance with the filling of the outermost shell of electrons, the so-called valency electrons. The inert gases each had completed outer valency shells with a full complement of eight electrons, and this made them virtually unreactive. The alkali metals, in Group I, had only one electron in their outermost shell, and were intensely avid to get rid of this, to attain the stability of an inert-gas configuration; the halogens in Group VII, conversely, with seven electrons in their valency shell, were avid to acquire an extra electron and also achieve an inert-gas configuration. Thus when sodium came into contact with chlorine, there would be an immediate (indeed explosive) union, each sodium atom donating its extra electron, and each chlorine atom happily receiving it, both becoming ionized in the process.

The placement of the transition elements and the rare-earth elements in the periodic table had always given rise to special problems. Bohr now suggested an elegant and ingenious solution to this: the transition elements, he proposed, contained an additional shell of ten electrons each; the rare-earth elements an additional shell of fourteen. These inner shells, deeply buried in the case of the rare-earth elements, did not affect chemical character in nearly so extreme a way as the outer shells; hence the relative similarity of all the transition elements and the extreme similarity of all the rare-earth elements.

Bohr's electronic periodic table, based on atomic structure, was essentially the same as Mendeleev's empirical one based on chemical reactivity (and all but identical with the block tables devised in pre-electronic times, such as Thomsen's pyramidal table and Werner's ultralong table of 1905). Whether one inferred the periodic table from the chemical properties of the elements or from the electronic shells of their atoms, one arrived at exactly the same point.[2] Moseley and Bohr had made it absolutely clear that the periodic table was based on a fundamental numerical series that determined the number of elements in each period: two in the first period, eight each in the second and third, eighteen each in the fourth and fifth; thirty-two in the sixth and perhaps also the seventh. I repeated this series—2, 8, 8, 18, 18, 32—over and over to myself.

At this point I started to revisit the Science Museum and spend hours once again gazing at the giant periodic table there, this time concentrating on the atomic numbers inscribed in each cubicle in red. I would look at vanadium, for example—there was a shining nugget in its pigeonhole—and think of it as element 23, a 23 consisting of 5 + 18: five electrons in an outer shell around an argon "core" of eighteen. Five electrons—hence its maximum valency of 5; but three of these formed an incomplete inner shell, and it was such an incomplete shell, I had now learned, that gave rise to vanadium's characteristic colors and magnetic susceptibilities. This sense of the quantitative did not replace the concrete, the phenomenal sense of vanadium but heightened it, because I saw it now as a revelation, in atomic terms, of why vanadium had the properties it did. The qualita-

---

[2] This gave Bohr predictive power too. Moseley had observed that element 72 was missing, but could not say whether it would be a rare-earth element or not (elements 57–71 were rare earths, and 73, tantalum, was a transition element, but no one was sure how many rare earths there would be). Bohr, with his clear idea of the numbers of electrons in each shell, was able to predict that element 72 would not be a rare-earth element, but a heavier analog of zirconium. He suggested that his colleagues in Denmark seek this new element in zirconium ores, and it was swiftly found (and named hafnium, after the old name for Copenhagen). This was the first time the existence and properties of an element were predicted not by chemical analogy, but on the purely theoretical basis of its electronic structure.

tive and the quantitative had fused in my mind; the sense of "vanadiumness" now could be approached from either end.

Between them, Bohr and Moseley had restored arithmetic to me, provided the essential, transparent arithmetic of the periodic table which had been intimated, though only in a muddy way, by atomic weights. The character and identity of the elements, much of it, anyhow, could now be inferred from their atomic numbers, which no longer just indicated nuclear charge but stood for the very architecture of each atom. It was all divinely beautiful, logical, simple, economical, God's abacus at work.

What made metals metallic? Electronic structure explained why the metallic state seemed to be fundamental, so different in character from any other. Some of the mechanical properties of metals, their high densities and melting points, could now be explained in terms of the tightness with which electrons were bound to the nucleus. A very tightly bound atom, with a high "binding energy," seemed to go with unusual hardness and density, and high melting point. Thus it was that my favorite metals—tantalum, tungsten, rhenium, osmium: the filament metals—had the highest binding energies of any of the elements. (So there was, I was pleased to learn, an atomic justification for their exceptional qualities—and for my own preference.) The conductivity of metals was ascribed to a "gas" of free and mobile electrons, easily detached from their parent atoms—this explained why an electric field could draw a current of mobile electrons through a wire. Such an ocean of free electrons, on the surface of a metal, could also explain its special luster, for oscillating violently with the impact of light, these would scatter or reflect any light back on its own path.

The electron-gas theory carried the further implication that under extreme conditions of temperature and pressure, all the nonmetallic elements, all matter, could be brought into a metallic state. This had already been achieved with phosphorus in the

1920s, and it was predicted, in the 1930s, that at pressures in excess of a million atmospheres it might be achieved with hydrogen, too—there might be metallic hydrogen, it was speculated, at the heart of gas giants like Jupiter. The idea that *everything* could be "metallized" I found deeply satisfying.[3]

I had long been puzzled by the peculiar powers of blue or violet light, short-wavelength light, as opposed to red or long-wavelength light. This was clear in the darkroom: one could have quite a bright ruby safelight that would not fog a developing film, whereas the least hint of white light, daylight (which of course contained blue), would fog it straightaway. It was clear, too, in the lab, where chlorine, for example, could be safely mixed with hydrogen in red light, but the mixture would explode in the presence of the least white light. And it was clear with Uncle Dave's mineral cabinet, where one could induce phosphorescence or fluorescence with blue or violet light, but not with red or orange light. Finally, there were the photoelectric cells that Uncle Abe had in his house; these could be activated by the merest pencil of blue light, but would not respond to even a flood of red light. How could a huge amount of red light be less effective than a tiny amount of blue light? It was only after I had learned something of Bohr and Planck that I realized the answer to these apparent paradoxes must lie in the quantal nature of radiation and light, and the quantal states of the atom. Light or radiation came in minimum units or quanta, the energy of which

---

[3] It was also wondered, early in the twentieth century, what might happen to the "electron gas" in metals if they were cooled to temperatures near absolute zero—would this "freeze" all the electrons, turning the metal into a complete insulator? What was found, using mercury, was the complete opposite: the mercury became a perfect conductor, a superconductor, suddenly losing all its resistance at 4.2 degrees above absolute zero. Thus one could have a ring of mercury, cooled by liquid helium, with an electrical current flowing around it with no diminution, for days, forever.

depended on their frequency. A quantum of short-wavelength light—a blue quantum, so to speak—had more energy than a red one, and a quantum of X-rays or gamma rays had far more energy still. Each type of atom or molecule—whether of a silver salt in a photographic emulsion, or of hydrogen or chlorine in the lab, or of cesium or selenium in Uncle Abe's photocells, or of calcium sulfide or tungstate in Uncle Dave's mineral cabinet—required a certain specific level of energy to elicit a response; and this might be achieved by even a single high-energy quantum, where it could not be evoked by a thousand low-energy ones.

As a child I thought that light had form and size, the flower-like shapes of candle flames, like unopened magnolias, the luminous polygons in my uncle's tungsten bulbs. It was only when Uncle Abe showed me his spinthariscope and I saw the individual sparkles in this that I started to realize that light, all light, came from atoms or molecules which had first been excited and then, returning to their ground state, relinquished their excess energy as visible radiation. With a heated solid, such as a white-hot filament, energies of many wavelengths were emitted; with an incandescent vapor, such as sodium in a sodium flame, only certain very specific wavelengths were emitted. (The blue light in a candle flame which had so fascinated me as a boy, I later learned, was generated by cooling dicarbon molecules as they emitted the energy they had absorbed when heated.)

But the sun, the stars, were like no lights on earth. They were of a brilliance, a whiteness, exceeding the hottest filament lamps (some, like Sirius, were almost blue). One could infer, from the radiation energy of the sun, a surface temperature of about 6,000 degrees. No one in his youth, Uncle Abe reminded me, had any idea what could allow the enormous incandescence and energy of the sun. *Incandescence* was scarcely the right word, for there was no burning, no combustion, in the ordinary sense—most chemical reactions, indeed, ceased above 1,000 degrees.

Could gravitational energy, the energy generated by a gigantic mass contracting, keep the sun going? This, too, it seemed, would be wholly inadequate to account for the blazing heat and energy of the sun and stars, undimmed for billions of years. Nor was radioactivity a plausible source of energy, because radioactive elements were not present in the stars in anywhere near the needed quantities, and their output of energy was too slow and unhurryable.

It was not until 1929 that another idea was put forth: the notion that, given the prodigious temperatures and pressures of a star's interior, atoms of light elements might fuse together to form heavier atoms—that atoms of hydrogen, as a start, could fuse to form helium; that the source of cosmic energy, in a word, was thermonuclear. Huge amounts of energy had to be pumped into light nuclei to make them fuse together, but once fusion was achieved, even more energy would be given out. This would in turn heat up and fuse other light nuclei, producing yet more energy, and this would keep the thermonuclear reaction going. The inside of the sun reaches enormous temperatures, something on the order of twenty million degrees. I found it difficult to imagine a temperature like this—a stove at this temperature (George Gamow wrote in *The Birth and Death of the Sun*) would destroy everything around it for hundreds of miles.

At temperatures and pressures like this, atomic nuclei— naked, stripped of their electrons—would be rushing around at tremendous speed (the average energy of their thermal motion would be similar to that of alpha particles) and continually crashing, uncushioned, into one another, fusing to form the nuclei of heavier elements.

> We must imagine the interior of the Sun [Gamow wrote] as some gigantic kind of natural alchemical laboratory where the transformation of various elements into one another takes place almost as easily as do the ordinary chemical reactions in our terrestrial laboratories.

Converting hydrogen to helium produced a vast amount of heat and light, for the mass of the helium atom was slightly less than that of four hydrogen atoms—and this small difference in mass was totally transformed into energy, in accordance with Einstein's famous $e = mc^2$. To produce the energy generated in the sun, hundreds of millions of tons of hydrogen had to be converted to helium each second, but the sun is composed predominantly of hydrogen, and so vast is its mass that only a small fraction of it has been consumed in the earth's lifetime. If the rate of fusion were to decline, then the sun would contract and heat up, restoring the rate of fusion; if the rate of fusion were to become too great, the sun would expand and cool down, slowing it. Thus, as Gamow put it, the sun represented "the most ingenious, and perhaps the only possible, type of 'nuclear machine,'" a self-regulating furnace in which the explosive force of nuclear fusion was perfectly balanced by the force of gravitation. The fusion of hydrogen to helium not only provided a vast amount of energy, but also created a new element in the world. And helium atoms, given enough heat, could be fused to make heavier elements, and these elements, in turn, to make heavier elements still.

Thus, by a thrilling convergence, two ancient problems were solved at the same time: the shining of stars, and the creation of the elements. Bohr had imagined an *aufbau,* a building up of all the elements starting from hydrogen, as a purely theoretical construct—but such an *aufbau* was realized in the stars. Hydrogen, element 1, was not only the fuel of the universe, it was the ultimate building block of the universe, the primordial atom, as Prout had thought back in 1815. This seemed very elegant, very satisfying, that all one needed to start with was the first, the simplest of atoms.[4]

---

[4] The universe started, Gamow conceived, as almost infinitely dense—perhaps no larger than a fist. Gamow and his student Ralph Alpher went on to suggest (in a famous 1948 article that came to be known, after Hans Bethe was

Bohr's atom seemed to me ineffably, transcendently beautiful—electrons spinning, trillions of times a second, spinning forever in predestined orbits, a true perpetual-motion machine made possible by the irreducibility of the quantum, and the fact that the spinning electron expended no energy, did no work. And more complex atoms were more beautiful still, for they had dozens of electrons weaving separate paths, but organized, like tiny onions, in shells and subshells. They seemed to me not merely beautiful, these gossamer but indestructible things, but perfect in their way, as perfect as equations (which indeed could express them) in their balancing of numbers and forces and shieldings and energies. And nothing, no ordinary agency, could upset their perfections. Bohr's atoms were surely close to Leibniz's optimum world.

"God thinks in numbers," Auntie Len used to say. "Numbers are the way the world is put together." This thought had never left me, and now it seemed to embrace the whole physical world. I had started to read a little philosophy at this point, and Leibniz,

---

invited to add his name, as the alpha-beta-gamma paper), that this primal fist-sized universe exploded, inaugurating space and time, and that in this explosion (which Hoyle, derisively, was to call the Big Bang) all of the elements were created.

But here he was wrong; it was only the lightest elements—hydrogen and helium and perhaps a little lithium—that originated in the Big Bang. It was not until the 1950s that it became clear how the heavier elements were generated. It might take billions of years for an average star to consume all its hydrogen, but the more massive stars, far from extinguishing at this point, could contract, becoming hotter still, and start on further nuclear reactions, fusing their helium to produce carbon, fusing this in turn to produce oxygen, and then silicon, phosphorus, sulfur, sodium, magnesium—all the way up to iron. Beyond iron no energy could be released by further fusion, so this accumulated as an end point in nucleosynthesis. Hence its remarkable abundance in the universe, an abundance reflected in metallic meteorites and in the iron core of the earth. (The heavier elements, those beyond iron, remained a puzzle for longer; they only originate, apparently, with supernova explosions.)

so far as I could understand him, appealed to me especially. He spoke of a "Divine mathematics," with which one could create the richest possible reality by the most economical means, and this, it now seemed to me, was everywhere apparent: in the beautiful economy by which millions of compounds could be made from a few dozen elements, and the hundred-odd elements from hydrogen itself; the economy by which the whole range of atoms was composed from two or three particles; and in the way that their stability and identity were guaranteed by the quantal numbers of the atom itself—all this was beautiful enough to be the work of God.

25

THE END OF THE AFFAIR

It was "understood," by the time I was fourteen, that I was going to be a doctor; my parents were doctors, my brothers in medical school. My parents had been tolerant, even pleased, with my early interests in science, but now, they seemed to feel, the time for play was over. One incident stays clearly in my mind. It was 1947, a couple of summers after the war, and I was with my parents in our new Humber touring the South of France. Sitting in the back, I was talking about thallium, rattling on and on and on about it: how it was discovered, along with indium, in the 1860s, by the brilliantly colored green line in its spectrum; how some of its salts, when dissolved, could form solutions nearly five times as dense as water; how thallium indeed was the platypus of the elements, with paradoxical qualities that had caused uncertainty about its proper placement in the periodic table—soft, heavy, and fusible like lead, chemically akin to gallium and indium, but with dark oxides like those of manganese and iron, and colorless sulfates like those of sodium and potassium. Thallium salts, like silver salts, were sensitive to light—one could have a whole photography based on thallium! The thallous ion, I continued, had great similarities to the potassium ion—similarities which were fascinating in the laboratory or factory, but utterly deadly to the organism, for, biologically almost indistinguishable from potassium, thallium would slip into all the roles and pathways of potassium, and sabotage the now-helpless organism from within. As I babbled on, gaily, narcissistically, blindly, I did not notice that my parents, in the front seat, had fallen completely silent, their faces bored, tight, and disapproving—until, after twenty minutes, they could bear it no longer, and my father burst out violently: "Enough about thallium!"

But it was not sudden—I did not wake up one morning and find that chemistry was dead for me; it was gradual, it stole upon me bit by bit. It happened at first, I think, without my even realizing it. It came upon me sometime in my fifteenth year that I no longer woke up with sudden excitements—"Today I will get the Clerici solution! Today I will read about Humphry Davy and electric fish! Today I will finally understand diamagnetism, perhaps!" I no longer seemed to get these sudden illuminations, these epiphanies, those excitements which Flaubert (whom I was now reading) called "erections of the mind." Erections of the body, yes, this was a new, exotic part of life—but those sudden raptures of the mind, those sudden landscapes of glory and illumination, seemed to have deserted or abandoned me. Or had I, in fact, abandoned them? For I was no longer going to my little lab; I only realized this when I wandered down one day and saw a light layer of dust on everything there. I had scarcely seen Uncle Dave or Uncle Abe for months, and I had ceased to carry my pocket spectroscope with me.

There had been times when I would sit in the Science Library, entranced for hours, totally oblivious to the passage of time. There were times when I seemed to *see* "lines of force" or electrons dancing, hovering, in their orbitals, but now this half-hallucinatory power was gone too. I was less dreamy, more focused, school reports said—that, perhaps, was the impression I gave—but what I felt was wholly different; I felt that an inner world had died and been taken from me.

I often thought of Wells's story about the door in the wall, the magic garden the little boy gets admitted to, and his subsequent exile or expulsion from it. He does not notice at first, in the press of life and outer achievement, that he has lost something, then the consciousness of this begins to grow on him, eroding and finally destroying him. Boyle had called his lab an "Elysium"; Hertz had spoken of physics as "an enchanted fairyland." I felt I was now outside this Elysium, that the doors of the fairyland were now closed to me, that I had been expelled from the garden of numbers, the garden of Mendeleev, the magic play realms to which I had had admittance as a boy.

With the "new" quantum mechanics, developed in the mid-1920s, one could no longer see electrons as little particles in orbit, one had to see them now as waves; one could no longer speak of an electron's position, only of its "wave function," the probability of finding it in a particular place. One could not measure its position and velocity simultaneously. An electron, it seemed, could be (in some sense) everywhere and nowhere at once. All this set my mind reeling. I had looked to chemistry, to science, to provide order and certainty, and now suddenly this was gone.[1] I found myself in a state of

---

[1] This question again resonated for me when I read Primo Levi's wonderful book *The Periodic Table,* especially the chapter called "Potassium." Here Levi speaks of his own search, as a student, for "sources of certainty." Deciding he would become a physicist, Levi left the chemistry lab and apprenticed himself to the physics department—to an astrophysicist, in particular. This did not work out quite as he had hoped, for while some ultimate certainties might indeed be found in stellar physics, such certainties, though sublime, were

shock, and I was past my uncles now, and in deep water, alone.[2]

This new quantum mechanics promised to explain all of chemistry. And though I felt an exuberance at this, I felt a certain threat, too. "Chemistry," wrote Crookes, "will be established upon an entirely new basis. . . . We shall be set free from the need for experiment, knowing a priori what the result of each and every experiment must be." I was not sure I liked the sound of this. Did this mean that chemists of the future (if they existed) would never actually need to handle a chemical; might never see

---

abstract and remote from daily life. More soul-filling, nearer life, were the beauties of practical chemistry. "When I understand what's going on inside a retort," Levi once remarked, "I'm happier. I've extended my knowledge a little bit more. I haven't understood truth or reality. I've just reconstructed a segment, a little segment of the world. That's already a big victory inside a factory laboratory."

[2] I was not quite alone. A most important guide to me at this point was George Gamow, a scientist-writer of great versatility and charm whose *Birth and Death of the Sun* I had already read. In his "Mr. Tompkins" books (*Mr. Tompkins in Wonderland* and *Mr. Tompkins Explores the Atom,* published in 1945), Gamow uses the device of altering physical constants by many orders of magnitude to make otherwise unimaginable worlds at least half-imaginable. Relativity is made comically imaginable by supposing the velocity of light to be only thirty miles per hour, and quantum mechanics equally so by imagining Planck's constant increased by twenty-eight orders of magnitude, so that one can have quantum effects in "real" life—thus quantum tigers, smeared out in a quantum jungle, are nowhere and everywhere at once.

I sometimes wondered whether any "macroquantal" phenomena existed, whether one might ever be able to see, under extraordinary conditions, a quantal world with one's own eyes. One of the unforgettable experiences of my life was exactly this, when I was introduced to liquid helium, and saw how this changed its properties suddenly at a critical temperature, turning from a normal liquid into a strange superfluid with no viscosity, no entropy whatever, able to go through walls, to climb out of a beaker, and with a thermal conductivity three million times that of normal liquid helium. This impossible state of matter could only be understood in terms of quantum mechanics: the atoms were now so close together that their wave functions overlapped and merged, so that one had, in effect, a single giant atom.

the colors of vanadium salts, never smell a hydrogen selenide, never admire the form of a crystal; might live in a colorless, scentless mathematical world? This, for me, seemed an awful prospect, for *I,* at least, needed to smell and touch and feel, to place myself, my senses, in the middle of the perceptual world.[3]

I had dreamed of becoming a chemist, but the chemistry that really stirred me was the lovingly detailed, naturalistic, descriptive chemistry of the nineteenth century, not the new chemistry of the quantum age. Chemistry as I knew it, the chemistry I loved, was either finished or changing its character, advancing beyond me (or so I thought at the time). I felt I had come to the end of the road, the end of my road, at least, that I had taken my journey into chemistry as far as I could.

I had been living (it seems to me in retrospect) in a sort of sweet interlude, having left behind the horrors and fears of Braefield. I had been guided to a region of order, and a passion for science, by two very wise, affectionate, and understanding uncles. My parents had been supportive and trusting, had allowed me to put a lab together and follow my own whims. School, mercifully, had been largely indifferent to what I was doing—I did my schoolwork, and was otherwise left to my own devices. Perhaps, too, there was a biological respite, the special calm of latency.

But now all this had changed: other interests were crowding in, exciting me, seducing me, pulling me in different ways. Life

---

[3] I wish I had realized—but that would not have been easy for me as a boy—that Crookes was wrong, that the new insight about the atom which prompted his thoughts (he was writing this in 1915, just two years after Bohr) would serve, once assimilated, to expand and enrich chemistry enormously, not to reduce it, annihilate it, as he feared. There were similar anxieties about the first atomic theory: many chemists, Humphry Davy among them, felt there was danger in accepting Dalton's notions of atoms and atomic weights, danger of pulling chemistry away from its concreteness and reality into an arid, impoverished, metaphysical realm.

had become broader, richer, in a way, but it was also shallower, too. That calm deep center, my former passion, was no longer there. Adolescence had rushed upon me, like a typhoon, buffeting me with insatiable longings. At school I had left the undemanding classics "side," and moved to the pressured science side instead. I had been spoiled, in a sense, by my two uncles, and the freedom and spontaneity of my apprenticeship. Now, at school, I was forced to sit in classes, to take notes and exams, to use textbooks that were flat, impersonal, deadly. What had been fun, delight, when I did it in my own way became an aversion, an ordeal, when I had to do it to order. What had been a holy subject for me, full of poetry, was being rendered prosaic, profane.

Was it, then, the end of chemistry? My own intellectual limitations? Adolescence? School? Was it the inevitable course, the natural history, of enthusiasm, that it burns hotly, brightly, like a star, for a while, and then, exhausting itself, gutters out, is gone? Was it that I had found, at least in the physical world and in physical science, the sense of stability and order I so desperately needed, so that I could now relax, feel less obsessed, move on? Or was it, perhaps, more simply, that I was growing up, and that "growing up" makes one forget the lyrical, mystical perceptions of childhood, the glory and the freshness of which Wordsworth wrote, so that they fade into the light of common day?

AFTERWORD

Toward the end of 1997, Roald Hoffmann—we had been friends since I had read his *Chemistry Imagined* a few years before—knowing something of my chemical boyhood, sent me an intriguing parcel. It contained a large poster of the periodic table with photographs of each element; a chemical catalog, so I could order a few things; and a little bar of a very dense, greyish metal, which fell onto the floor as I opened the package, landing with a resonant clonk. I recognized it at once by its feel and its sound ("the sound of sintered tungsten," my uncle used to say, "nothing like it").

The clonk served as a sort of Proustian mnemonic, and instantly brought Uncle Tungsten to mind, sitting in his lab in his wing collar, his shirtsleeves rolled up, his hands black from powdered tungsten. Other pictures rose immediately in my mind: his factory where the lightbulbs were made, his collections of old lightbulbs, and heavy metals, and minerals. And my own initiation by him, when I was ten, into the wonders of metallurgy and chemistry. I thought I might write a brief sketch of Uncle Tungsten, but the memories, now started, continued to emerge—memories not just of Uncle Tungsten but of all the events of early life, of my boyhood, many forgotten for fifty years or more. What had started as a page of writing became a vast mining operation, a four-year excavation of two million words or more—from which, somehow, a book began to crystallize out.

I have pulled out my old books (and bought many new ones), set the little tungsten bar on a tiny pedestal, and papered the kitchen with chemical charts. I read lists of cosmic abundances in the bath. On cold, dismal Saturday afternoons, I may curl up

with a fat volume of Thorpe's *Dictionary of Applied Chemistry*—
it was one of Uncle Tungsten's favorite books—opening it
anywhere and reading at random.

The passion for chemistry, which I had thought dead at fourteen, has clearly survived, deep inside me, throughout the inter-
vening years. Though my life has taken a different direction, I
have followed the new discoveries in chemistry with excitement.
In my day, elements stopped with number 92, uranium, but I
have watched closely as new elements—elements up to 118!—
have been made. These new elements probably exist only in the
lab and do not occur anywhere else in the universe, but I was
delighted to learn that the very latest of them, though still
radioactive, are thought to belong to a long-sought "island of stability," in which the atomic nuclei are almost a million times
more stable than those of the preceding elements.

Astronomers now wonder about giant planets with cores of
metallic hydrogen, stars made of diamond, and stars with crusts
of iron helide. The inert gases have been coaxed into combination, and I have seen fluorides of xenon—almost unthinkable, a
fantasy for me, in the 1940s. The rare-earth elements, which
both Uncle Tungsten and Uncle Abe so loved, have now become
common and find countless uses in fluorescent materials, phosphors of every color, high-temperature superconductors, and tiny
magnets of an unbelievable strength. The powers of synthetic
chemistry have become prodigious: we can design molecules now
with almost any structure, any property, we wish.

Tungsten, with its density and hardness, has found new uses in
darts and tennis rackets and—disturbingly—in coating shells
and missiles. But it also turns out—this is much more to my
taste—to be indispensable to certain primitive bacteria which
get their energy by metabolizing sulfur compounds in the
hydrothermal vents of the ocean depths. If (as is now speculated)
such bacteria were the first organisms on earth, then tungsten
may have been crucial for the origin of life.

The old enthusiasm surfaces every so often in odd associations and impulses: a sudden desire for a ball of cadmium, or to feel the coldness of diamond against my face. The license plates of cars immediately suggest elements, especially in New York, where so many of them begin with U, V, W, and Y—that is, uranium, vanadium, tungsten, and yttrium. It is an added pleasure, a bonus, a grace, if the symbol of an element is followed by its atomic number, as in W74 or Y39. Flowers, too, bring elements to mind: the color of lilacs in spring for me is that of divalent vanadium. Radishes, for me, evoke the smell of selenium.

Lights—the old family passion—continue to evolve in wonderful ways. Sodium lights, a yellow glory, became widespread in the 1950s, and quartz-iodine lights, blazing halogen lamps, came out in the 1960s. If I wandered with a pocket spectroscope as a twelve-year-old in Piccadilly after the war, I have rediscovered the same joy now, walking with a pocket spectroscope through Times Square, seeing the city lights of New York as atomic emissions.

And I often dream of chemistry at night, dreams that conflate the past and the present, the grid of the periodic table transformed to the grid of Manhattan. The location of tungsten, at the intersection of Group VI and Period 6, becomes synonymous here with the intersection of Sixth Avenue and Sixth Street. (There is no such intersection in New York, of course, but it exists, conspicuously, in the New York of my dreams.) I dream of eating hamburgers made of scandium. Sometimes, too, I dream of the indecipherable language of tin (a confused memory, perhaps, of its plaintive "cry"). But my favorite dream is of going to the opera (I am Hafnium), sharing a box at the Met with the other heavy transition metals—my old and valued friends—Tantalum, Rhenium, Osmium, Iridium, Platinum, Gold, and Tungsten.

ACKNOWLEDGMENTS

I owe a huge debt to my brothers, my cousins, and, not least, my old friends, who have shared memories, letters, photographs, and memorabilia of all kinds; I could not have reconstructed the events of so long ago without them. I have written of them, and others, with some trepidation: "It is always dangerous," as Primo Levi remarked, "transforming a person into a character."

Kate Edgar, my assistant, and editor of many of my previous books, has been a virtual collaborator on this one, not only editing the innumerable drafts I produced, but meeting chemists with me, going down mines, enduring smells and explosions, electrical discharges and occasional radioactive emanations, and putting up with an office increasingly filled with periodic tables, spectroscopes, crystals dangling in supersaturated solutions, coils of wire, batteries, chemicals, and minerals. This book would still be a two-million-word excavation had it not been for her powers of distillation.

Sheryl Carter, also working with me, has opened the wonders of the Internet for me (I am computer-illiterate, and I do all my writing with a pen or an old typewriter), and has found books and articles and scientific instruments and toys of all sorts which I could never have got for myself.

In 1993, I wrote an essay-review in the *New York Review of Books* of David Knight's book on Humphry Davy, which in many ways rekindled my long-dormant interest in chemistry. I am grateful to Bob Silvers for encouraging me in this.

My article "Brilliant Light," an early fragment of this book which appeared in *The New Yorker,* was brilliantly edited (and titled) by my editor there, John Bennet; and Dan Frank, at Knopf, has been crucial in helping to steer the book to its present form.

Soon after starting this book I had the great pleasure of meeting a boyhood hero, Glenn Seaborg, and I have subsequently met or corresponded with chemists all over the world. These chemists, too many to name, have been astonishingly hospitable to an outsider, an ex-boy-enthusiast, and have shown me wonders that the wildest science fiction of my boyhood could not have conceived, such as "seeing" actual atoms (through the tungsten tip of an atomic force microscope), as well as humoring some nostalgic desires to see, once again, among other things, the deep blue of sodium dissolved in liquid ammonia; and tiny magnets levitated over superconductors cooled in liquid nitrogen, the magical, gravity-defying floating I had dreamed of as a child.

But, above all, it has been Roald Hoffmann who has been infinitely stimulating and supportive, and who has done more than anyone else to show me the marvelous thing which chemistry is now — and it is to Roald, therefore, that I dedicate this book.

INDEX

luster of, 35, 39, 103, 124,
302
sounds made by, 3, 4, 8, 315,
317
see also elements, filament
metals, native metals
microscopy, 104n, 238, 320
migraine, visual alterations in,
143
Miller, Jonathan, 16, 95, 123,
271–4, 276–7
minerals, 57–60, 63–6, 75, 196
and mining, 36–8, 39, 61, 65,
106, 107–8n, 117, 129,
137
names of, 59–60
transparent, 60 and n
Moissan, Henri, 38n, 84
molecules, nature of, 153–4
molybdenum, 43, 44, 52, 63–4,
206
Monterey Bay, 278–9
Morveau, Guyton de, 108–9, 111
Moseley, Harry, 294–6, 299n,
301 and n
mother, *see* Sacks, Elsie Landau
motorcycles, 23, 95, 236
museums, 57–9, 64, 69, 117,
145, 151, 154, 191, 203n,
205, 211, 272, 301
music, 181–4, 190, 195, 196,
269–70, 276
atoms as musical, 82n,
219–20, 298
teachers, 182, 184, 269
see also piano

nannies, 33, 99
Napoleon, 148n
Napoleon III, 130
narcotics, 73
National Geographic, 133–4, 141
native metals, 35, 41
nature, as reassuring 21, 190, 203
neon, 202, 219, 231
neurology, 95, 280
Newlands, J. A., 201n
Newton, Isaac, 102, 105n, 109,
113, 118, 121n, 125, 127,
149 and n, 151, 214, 227,
265
niobium, 61, 64, 74, 206
nitrogen
oxides of, 85, 118–9
Nobel Prize, 52, 246, 259
Noddack, Ida Tacke, 205n
nomenclature, 112–13
nuclear fission and fusion, 205n,
290n, 305–6, 306n
see also energy, atomic
numbers
as fundamental organizing
principle, 28, 190, 219–20,
296, 301–2, 307–8
passion for, 32, 222
in periodic table, 190, 194 and
n, 301
prime, 26–7
refuge in, 26–7
see also atomic numbers

octopus, 233, 273
Odling, William, 201n

A NOTE ON THE TYPE

The text of this book was set in Garamond No. 3. It is not a true
copy of any of the designs of Claude Garamond (ca. 1480–1561),
but an adaptation of his types, which set the European standard for
two centuries. This particular version is based on an adaptation by
Morris Fuller Benton.

Composed by North Market Street Graphics,
Lancaster, Pennsylvania
Printed and bound by Quebecor Graphics,
Fairfield, Pennsylvania
Designed by Peter A. Andersen